Cook, Taste, Learn

Arts and Traditions of the Table: Perspectives on Culinary History

Cook, Taste, Learn

How the Evolution of Science Transformed the Art of Cooking

Guy Crosby

COLUMBIA UNIVERSITY PRESS
NEW YORK

Columbia University Press
Publishers Since 1893
New York Chichester, West Sussex
cup.columbia.edu
Copyright © 2019 Guy Crosby
All rights reserved

Library of Congress Cataloging-in-Publication Data
Names: Crosby, Guy, author.
Title: Cook, taste, learn : how the evolution of science transformed the
art of cooking / Guy Crosby.
Description: First edition. | New York : Columbia University Press, [2019] |
Includes bibliographical references and index.
Identifiers: LCCN 2019021926 | ISBN 9780231192927 (cloth)
Subjects: LCSH: Cooking—Technique—History. | Gastronomy—History. |
Chemistry, Technical—History.
Classification: LCC TX645 .C76 2019 | DDC 641.01/3—dc23
LC record available at https://lccn.loc.gov/2019021926

Columbia University Press books are printed on permanent
and durable acid-free paper.
Printed in the United States of America

Cover design: Milenda Nan Ok Lee
Cover and frontispiece art: Jehan Georges Vibert (French, 1840–1902). *The Marvelous Sauce*, ca. 1890.
Collection Albright-Knox Art Gallery, Buffalo, New York; Bequest of Elisabeth H. Gates, 1899 (1899:3.14).
Image courtesy Albright-Knox Art Gallery. The painting contrasts the traditional art of cooking with the
evolving science of cooking based on the scientific method of cooking, tasting, and learning.
Back cover (detail): Guy Crosby. *Peaches in a Glass Bowl*, 1956. Courtesy of the artist.

For Christine, Kristin, Justin, Grace, and Mike

Contents

Preface

It's pretty amazing when you think about it: of all the living species on earth, only humans cook their food! Cooking has played a very important role in the evolution of humans, but only in the last few decades has the public shown much interest in the science of what happens to food when we cook it by applying heat. What is it about cooking food that has been so important in the evolution of humans? Why have humans evolved biologically and socially so much faster than all other living species? What role, if any, has cooking played?

Plenty, if you ask Richard Wrangham, professor of biological anthropology at Harvard University. Wrangham makes a compelling case for the role of cooking in his popular book *Catching Fire: How Cooking Made Us Human*. He is not the first to argue that cooking influenced human evolution, but he is the first to claim that early humans began using fire to cook food as early as 1.8 to 1.9 million years ago. Others believe that physical evidence found in caves indicates fire was first used for cooking about 400,000 years ago. Wrangham finds that the evolutionary evidence of one of our earliest ancestors, *Homo erectus*—including the increasing size of the brain and the decreasing size of the teeth and gastrointestinal system—strongly supports the conclusion that the use of fire to cook food began more than a million years earlier. Cooking clearly played an extremely important role in human evolution. Through knowledge of cooking science, we can understand why cooking was so fundamental to the evolution of humans. Throughout this book, I will explore the evolution of science and how it has transformed the art of cooking right up to the present.

Unfortunately, science is a mystery to many people. Its complex language is composed of strange words that make it difficult to understand, inviting suspicion or lack of interest. Yet science is a proven method for questioning and learning how the world works. It leads to new insights and the creation of new ideas, knowledge, and innovations. Through carefully designed studies, hypotheses of why certain things happen can be tested and affirmed or disproved, following the scientific method of cooking, tasting, and learning. Consider, for example, the theory of gravity (Newton), proof of the concept of atoms (Dalton), the invention of the electric lightbulb (Edison), the development of a polio vaccine (Salk), the design of the electronic transistor (Bardeen, Brattain, and Shockley), and the answer to why the chicken crossed the road (Henny Penny). Some people are suspicious that science can be used for negative purposes such as the development of atomic weapons, the creation of genetically modified organisms, or electronic surveillance. And, of course, in the hands of dishonest, politically or financially motivated people, it can. But on the whole, science has had a very positive impact on the evolution of humankind. Despite the long history of scientific thought, beginning with the ancient Egyptians and early Greek philosophers, the science of cooking has received little attention until very recently. The science of cooking may seem to be of minor importance, but if Wrangham is correct, it may be one of the most consequential and least appreciated of all the sciences.

The famous American sociologist Robert K. Merton (1910–2003, Columbia University) once wrote, "Science is public, not private, knowledge." That's wonderful for those who understand the science. But most science communication is perceived as pure gibberish. For some, it can be more opaque than even private knowledge. As a scientist, I feel it is essential to interpret science for the public so it can truly become public knowledge appreciated and understood by all. To explain the science of cooking, I have included a series of short essays on the cream-colored pages that delve deeper into the science considered in each chapter, as well as a small number of my favorite recipes illustrating that science. It is my hope that the growing interest in these topics will lead to a greater appreciation and understanding of science and our world.

I close this brief preface with my favorite lines of poetry:

> To see a World in a Grain of Sand
> And a Heaven in a Wild Flower
> Hold Infinity in the palm of your hand
> And Eternity in an hour.

These are the opening lines to "Auguries of Innocence" by William Blake (1757–1827), assumed to have been written in 1803 but not published until 60 years later. I first heard these lines spoken by Jacob Bronowski (1908–1974) in 1973 during his very successful BBC television documentary series *The Ascent of Man*. I have cherished them ever since. For me, these lines eloquently express the concept of the atom and the structure of matter built upon atoms. Yet, incredibly, the first precise proof of the modern concept of atomic theory was not published until 1805 by John Dalton, a Quaker schoolmaster in Manchester, England—2 years after Blake wrote his poem. How could Blake have imagined that grains of sand and wildflowers are composed of infinitesimally small molecular structures composed of even smaller atoms before the existence of the atom was proved? The answer will become apparent later in this book. The history of science and our understanding of the world are fascinating. I hope you find some of the same fascination and joy in the science of cooking that follows.

Acknowledgments

The inspiration for this book emanated from *Catching Fire: How Cooking Made Us Human*, written by Professor Richard Wrangham of Harvard University. I also wish to thank Professor Gordon Shepherd of Yale University for recommending me to the publisher, Columbia University Press. There are many who helped guide me through the publication process, including Jennifer Crewe, Associate Provost and Director at Columbia University Press, and her staff, including Monique Briones, Lisa Hamm, Milenda Lee, Marielle Poss, Meredith Howard, Justine Evans, Brian Smith, and Patrick Fitzgerald, as well as Ben Kolstad, Editorial Services Manager, and Sherry Goldbecker at Cenveo Publishing Services. I am also grateful for the work of Katherine Langenberg and Sabin Orr, who provided photographs for some of the illustrations, as well as many people who helped provide permission for use of other illustrations including Jeanne Brewster, Nadine Sobusa, Justin Hobson, Laurie Lounsberry McFadden, Dan Sousa, and Ian Matzen. Help with some of the science, writing, and reviewing was generously provided by Pia Sorensen, Adriana Fabbri, Nicoletta Pellegrini, Ali Bouzari, and Rebecca Dowgiert. In the early stages of writing this book I received very helpful guidance from Zick Rubin, James Levine, and Danielle Svetcov. I would also like to thank Christopher Kimball and Jack Bishop for providing me the opportunity to coauthor my first two books on cooking science published by America's Test Kitchen. I owe a special thank you to Professor Walter Willett for giving me the opportunity to teach a course on food science and technology at the Harvard T. H. Chan School of Public Health, which allowed me to gather much of the information for this book, as well as to Professor Carol Russell who made it possible for me to teach at Framingham State University following my retirement. Thanks also to

Harold McGee for catalyzing the public's interest in cooking science. Finally I would like to acknowledge the friendship, inspiration, and mutual interest in food, wine, and cooking provided by our close friends Ruth and Bill Benz, Margaret and Rick Perez, Sandy and Jerry Peters, Sue and Ron Daniels, and Darice and Bob Wareham. Thank you everyone for helping to make this book possible.

Cook, Taste, Learn

I

The Evolution of Cooking (2 Million–12,000 Years Ago)

Fire, Cooking, and the Evolution of Humans

Why are humans the only species living on earth that cooks its food? No one really knows for sure. There is solid archeological evidence that early humans were using fire for cooking about 400,000 years ago. Yet Richard Wrangham, an anthropologist at Harvard University, believes that biological evidence, such as the increasing size of the brain and decreasing size of the teeth and digestive system, suggests cooking may have started much earlier—almost 2 million years ago. He hypothesizes that early humans such as *Homo erectus* obtained more energy and nutrients from food through cooking and thus developed bigger brains, which gave them an evolutionary advantage over all other species. He also suggests that the key evolutionary changes may have occurred over a relatively short time span, less than 100,000 years, giving *Homo erectus* a competitive jump start. To be fair, a number of scientists disagree with Wrangham's conclusions, finding them to be unsupported by hard evidence. Whether or not we accept Wrangham's hypothesis that cooking food started almost two million years ago, cooking did offer other advantages. It made food safer to eat by killing pathogenic microorganisms, and it imparted a different, very appealing flavor that encouraged repetition of the dangerous act of cooking with fire. Ultimately, early humans began to evolve into more advanced present-day *Homo sapiens* about 195,000 years ago, by which time the volume of the human brain had increased by nearly 60 percent over that of the earlier *Homo erectus*. Was this change a result of early humans cooking their food?

FIGURE 1.1

Rock art in the Tassili n'Ajjer cave in Algeria, estimated to be 5,000–6,000 years old, depicts early humans gathered around an extinguished fire. Photograph attributed to Patrick Gruban/Wikimedia.

Almost certainly, early humans encountered naturally occurring fires well before they started cooking, and they probably perceived these fires as a threat to be feared. But the onset of the ice age during the Pleistocene era, which lasted from about 2.6 million to 11,700 years ago, provided the impetus to create, use, and control fire for warmth, light, and safety from predatory animals. Fire may also have had profound social consequences by promoting gathering and trade between very small groups of early humans. Cooking likely first happened by accident, perhaps when a wild boar was burned by natural fire or a small animal caught by hunters was left too near the open flames of a human-made fire. The alluring flavor of food roasted by fire likely led early humans to repeat the act again and again, as many recent studies have proven that, even today, flavor is the single most important factor in determining the foods we like and choose to eat. Limited contact between the disparate small groups of early humans may have slowed the spread of cooking beyond the first adapters, and thousands of years passed before cooking became a widespread practice.

The Taste, Smell, and Flavor of Food

Even today people enjoy the flavor of food cooked over open flames or hot coals, along with the sense that grilling represents a return to a much earlier way of cooking. The abilities to smell and taste predate the earliest humans by billions of years. Examinations of the ancient ancestors of present-day fish and amphibians indicate that specialized organs for sensing appeared well before these species migrated onto land. Both taste and smell appear to have evolved for survival and may be among the earliest senses developed, although not all modern-day species react to the same tastes and smells. For example, cats cannot detect sweet taste but are more sensitive to umami taste than humans—perhaps because cats evolved on a diet rich in animal protein. This variety suggests that different species evolved over time to detect the specific tastes and smells that were most critical for their survival. Today about 25 percent of humans are much more sensitive to bitter taste, with the remainder having either average or much less sensitivity. You might think these so-called supertasters would have a significant evolutionary advantage, as avoiding bitter-tasting toxic substances would make their survival more likely. But due to their heightened sensitivity to bitterness, supertasters are picky eaters and tend to avoid vegetables, many of which, like healthy broccoli and kale, are quite bitter. As a result, supertasters have a higher incidence of colon polyps, which may lead to colon cancer. Nontasters tend to like spicy food, as well as fat and alcohol, resulting in a greater tendency to be overweight, while average tasters tend to like most foods. When it comes to survival, this may be an example of when it is better to be average than above average.

Humans are hardwired in our DNA to detect taste from birth, or shortly after. We now have evidence that humans detect six basic tastes: sweet, salty, bitter, sour, umami, and fat. The tastes for sweet, bitter, umami, and fat are detected by protein receptors located on the surface of specific taste cells, while receptors for salty and sour are composed of ion channels within the cell membranes of taste cells. Much of the mouth cavity and the upper part of the tongue contain specialized receptor cells capable of detecting these tastes. Taste receptor cells contained within the larger taste buds are continuously replaced every 9–15 days due to repeated exposure to mechanical abrasion and very hot foods. Thus the taste buds are composed of about one hundred smaller individual taste cells and are part of the larger papillae that appear as the visible bumps on the tongue as well as throughout much of the mouth. In fact, the entire taste organ is one of only a few human organs capable of total regeneration.

Sweet taste is critical for detecting sugars, which are required for quick energy. The human brain uses the simple sugar glucose for almost all of its energy, requiring as much as 4.2 ounces (about 120 grams) per day to keep it functioning. Bitter taste evolved to guard against eating toxic plants, as many toxic substances taste very bitter. Humans have about twenty-five different types of receptors for bitter taste and only one type for sweet taste. Avoiding poisonous substances was apparently more important than detecting sweet foods needed for energy. Humans are about a thousand times more sensitive to bitter substances than sweet ones. We need to avoid even tiny amounts of toxic, bitter substances while consuming much more sugar for energy, so we have evolved different levels of sensitivity for each taste. Salt (sodium chloride) is critical for maintaining the fluid levels in our body, so the ability to detect salt is very important. Umami, described as a savory or meaty taste, probably evolved to detect sources of proteins and the essential amino acids that our body cannot make. Amino acids are used to build many important nutrients in our body, such as hormones, and DNA while also providing an additional source of energy. Fat, which is required for storing long-term energy, is also the source of two essential fatty acids that our body cannot make, linoleic acid and linolenic acid. Sour taste is a common property of all acids. Vitamin C, or ascorbic acid, is an essential nutrient that our body cannot make. Our sense of sour taste probably evolved as a means to detect possible sources of vitamin C.

HOW MANY BASIC TASTES ARE THERE?

Four basic tastes—sweet, salty, sour, and bitter—have been recognized as far back as the ancient Greeks and Romans. At one time, it was believed that each of the tastes was sensed in a different part of the mouth and tongue. But it has long been known that this simple picture is incorrect and that we sense tastes throughout the mouth, at the back of the throat, and on the top of the tongue. The concept of four basic tastes was rigorously adhered to until the early 1900s, when Kikunae Ikeda, a Japanese physical chemist at the University of Tokyo, decided to search for the taste of *konbu*, a seaweed that was commonly used to contribute a unique delicious, savory taste, which he named umami, to many Japanese dishes, such as the soup known as *konbu dashi*. In 1908, after a year of careful experimentation, Ikeda was able to isolate a very small amount of a pure chemical compound as the source of the umami taste in seaweed; he then proved the chemical structure was the same as that of the previously known sodium salt of glutamic acid, an important amino acid in the human diet. Interestingly, only salts (potassium, sodium, and calcium) of glutamic acid elicit the umami taste. Ikeda proclaimed that umami was the fifth basic taste because it was so commonly encountered in Japanese food. But this was not well received by other cultures around the world where seaweed was not commonly used in cooking, and for much of the twentieth century, umami was not widely accepted as the fifth basic taste.

In 1913, a student of Ikeda named Shinto Kodama undertook a study to identify the main source for the taste of bonito flakes, a form of dried tuna, which were also widely used to season Japanese food. He identified a type of chemical compound called a *nucleotide*, with the name inosine-5'-monophosphate (IMP), as a new component of the umami taste of bonito flakes. Further research found chemically similar nucleotides in shiitake mushrooms, another source of umami taste. In all, nucleotides responsible for umami taste were found in both animal and vegetable products, including tuna, dried sardines, beef, pork, chicken, Parmesan cheese, tomatoes, mushrooms, and fermented soybean products, such as soy sauce. Many years later, in 1967, it was recognized that combining the salts of glutamic acid and another amino acid named aspartic acid with the nucleotides increased the intensity of umami taste by as much as twenty times that of sodium glutamate or the nucleotides alone. This is an example of a true synergistic, or magnifying, effect. Thus combinations of foods such as tomato sauce with Parmesan cheese and beef with mushrooms have a much stronger savory umami taste than either food alone. This discovery has allowed cooks to greatly intensify the savory taste of dishes by combining certain ingredients that supply both sodium glutamate and nucleotides. With this discovery, umami began to be accepted more widely as the fifth basic taste. Ultimately, between 1998 and 2000, researchers used molecular biology to prove the existence of a specific protein receptor for sodium glutamate on taste receptor cells in the mouth, and with this discovery, umami was finally accepted as the fifth basic taste.

Aging and fermenting many protein-rich foods such as fish, meat, cheese, and soybeans produce both sodium glutamate and nucleotides, thus creating the source of intense umami taste found in fermented anchovies, aged Parmesan cheese, dry-aged beef, and fermented soy sauce. Today these ingredients are routinely used to heighten the savory taste of dishes around the world. Fermentation breaks down proteins into peptides and amino acids, the source of umami taste. In fact, glutamic acid and aspartic acid are the most abundant amino acids in legumes, wheat, meat, poultry, eggs, and dairy products such as cheese and milk from cows and humans. Fermentation is also a source of nucleotides, especially from fish such as anchovies. More than 2,000 years ago ancient Romans fermented small fish in the presence of salt for several months to produce a fish sauce called *garum*. The Romans were very fond of using *garum* to season many dishes, but they did not identify it as a fifth basic taste, primarily because they were rigid believers that there were only four basic tastes.

Until even more recently, it was widely believed that humans sense only five basic tastes. But acceptance that taste evolved as a mechanism for survival led many to believe there may be other tastes that we sense. The most logical is the sense of taste for fat, as fat is important not only for storing long-term energy but also for providing the two essential fatty acids required for a healthy diet, linoleic acid and linolenic acid, which occur in plant oils. In 2012, a gene labeled CD36, located on chromosome 7, was identified as the gene responsible for producing the taste receptor protein for sensing medium- and long-chain fatty acids in the mouth. Further well-designed taste tests published in 2015 by researchers at Purdue University confirmed that the taste of fat (actually free fatty acids) was indeed a sixth basic taste. They coined a new word, *oleogustus*, to mean the taste of fat. (*Oleo* is Latin for "fatty," and *gustus* means "taste.") In the future, we may discover even more basic tastes, but as of today, it is accepted that humans sense six basic tastes: sweet, salty, sour, bitter, umami, and fat.

REFERENCES

Kurihara, K. "Glutamate: From Discovery as a Food Flavor to Role as a Basic Taste (Umami)." *American Journal of Clinical Nutrition* 90, supp. (2009): 719S–22S.

Pepino, M., L. Love-Gregory, S. Klein, and N. Abumarad. "The Fatty Acid Translocase Gene CD36 and Lingual Lipase Influence Oral Sensitivity to Fat in Obese Subjects." *Journal of Lipid Research* 53 (2012): 561–566.

Running, C., B. Craig, and R. Mattes. "Oleogustus: The Unique Taste of Fat." *Chemical Senses* 40 (2015): 1–10. https://doi.org/10.1093/chemse/bjv036.

Interestingly, humans are far more sensitive to smell than taste. There are many molecules that most people can smell at the level of one part per trillion—and even at much lower levels. To put this in context, one part per trillion is equivalent to 1 second in 32,000 years! That's an awfully small amount. An example is the compound primarily responsible for the aroma of green bell peppers. One drop of this pure compound is enough to make a 30,000-gallon swimming pool smell like green bell pepper. Why are we so sensitive to smell? In terms of survival, the answer is not as obvious as it is for taste. Decomposing plant and animal proteins develop a strong odor from the formation of volatile compounds called *amines*. Think of the smell of old fish. So smell may have evolved to help humans avoid consuming rotting food, but more likely it evolved as a means of communicating through scent, allowing humans to detect minute amounts of pheromones, before language was developed. Our nose contains about four hundred different types of receptors for smell. In fact, the genes responsible for producing the protein receptors for smell comprise the largest family of genes in the human genome. We are capable of detecting far more than ten thousand different odors, which is a learned skill, unlike our ability to distinguish tastes, which is stored in our DNA. Humans detect smells that enter through the front of the nose, called *orthonasal smell*, as well as those that pass into the nose from the back of the mouth, called *retronasal smell*. Retronasal smell created by chewing and swallowing has been shown to be the most important contributor to the aroma and flavor of food.

Flavor is not something we sense with our nose or mouth like we do taste and smell. The sense of flavor is created in our brain from the electrical signals coming from our receptor cells for taste and smell. Thus, taste, smell, and flavor are distinct from each other. A number of different centers in the brain process the signals from our mouth and nose, creating an image of flavor in our mind. Our craving for certain favorite foods is created in three different sections of the brain, the same three sections that create our cravings for sex, addictive drugs, and music. Is there any wonder humans have such a strong desire for food, one that goes well beyond the need for energy and nutrition? These mental images of flavor include the comfort foods we learn to love as children and crave as adults. They are the source of the desire for fast foods like hamburgers, French fries, mac and cheese, and pizza that we first ate when much younger. Evidence suggests we develop more receptors for the comfort foods we crave, thus magnifying the effect over time. The craving for our favorite foods goes beyond the sensations of taste and smell and the images of flavor created in our brain. The texture of food, which is the physical feeling of the food in our mouth (think of slimy or chewy or crunchy); the sound when we chew it (think of crispy potato chips); its temperature; how appealing it looks; and, above all, the memories associated with eating it all combine to determine which foods we crave.

RECIPE 1.1 Linguine with Clam Sauce

The following recipe serves two (but see note at end):

INGREDIENTS:

12 small whole fresh clams in the shell

2 Tbsp. extra virgin olive oil

1 medium onion, chopped into very small pieces

2 garlic cloves, finely chopped

¼ cup dry vermouth

2 cans (6½ ounces, or 184 grams) cooked chopped clams in clam juice

¼ cup chopped fresh flat-leaf parsley

6 ounces dry linguine or spaghetti

1 Tbsp. fresh lemon juice

Parmigiano Reggiano cheese, freshly grated, to taste

NOTE:

Enough sauce can be made for four people by increasing the number of cans of chopped clams to three rather than two.

***Yield*: 2 servings**

Over the past several decades, I have made this recipe countless times, and each time it tastes just as delicious as it did the time before. The secret lies in the intense flavor of the clam sauce. Using canned chopped clams cooked in their juice not only makes this a simple, quick dish to prepare but also provides the remarkable flavor. The compound responsible for the intense taste and aroma holds the record for being detectable by humans at the lowest concentration recorded so far: 10–5 parts per trillion, or 0.01 parts per quadrillion! To put this in perspective, one part per quadrillion is equivalent to 2½ minutes out of the age of the earth (4.5 billion years)! For more explanation, read box 4.1, Numbers Both Large and Small. This compound's chemical name—that is, how an organic chemist defines the chemical structure of this compound—is pyrrolidinol-[1,2-e]-4H-2,4-dimethyl-1,3,5-ditiazine. Like so many sulfur-containing compounds, it has an intense aroma that is formed in trace amounts in boiled clams and shrimp. One whiff of an opened can of chopped clams in their juice will provide the unmistakable smell of the compound.

I like to embellish the dish by adding a modest number of small whole clams, such as count neck or Long Island clams. They are more for looks than anything else and do not significantly enhance the flavor, but they are fun to eat from the shells. Omit them if you can't find whole clams in your supermarket or you don't want to fuss with them.

DIRECTIONS:

Soak the whole clams in very cold water for about 1 hour and drain.

Heat a 10-inch stainless steel skillet on medium heat. Add the olive oil and onion, and gently cook, stirring occasionally, until the onion is soft but not browned, about 5 minutes. Add the garlic, and cook for an additional minute, stirring occasionally. Briefly remove the pan from the heat, and allow it to cool for about 1 minute (to prevent the vermouth from splattering in the hot pan). Add the vermouth, and continue cooking until most of the vermouth has evaporated. Add the chopped clams along with all of their juice, and cook the mixture vigorously until about 1/3 cup of liquid remains in the pan. Distribute the whole clams evenly in the pan along with the chopped parsley; cover and continue to cook until the shells of all the clams have opened (discard any clams that do not open). Remove the opened clams in their shells with tongs, and place them evenly around the perimeter of two plates (six clams per plate). Add freshly ground pepper to the clam sauce (the canned clams contain enough salt that extra is usually not needed), and reserve off heat.

While the sauce is being prepared, bring a large pot of lightly salted water to the boil, add the linguine, and cook until al dente. Reserve about 1/2 cup of the pasta water to add to the sauce if necessary. Strain the pasta, but do not rinse, and return it to the large pot; add the hot clam sauce (without the whole clams) to the pasta, and mix along with the lemon juice. Add a small amount of the hot pasta water if necessary to give the sauce the desired consistency; it should thoroughly coat the linguine but not be too soupy or too dry (about 1/3 cup of liquid is usually sufficient). Place the linguine and sauce mixture in the middle of the plates with the whole clams around the edges, and serve along with freshly grated Parmigiano Reggiano cheese.

The Impact of Cooking on the Human Brain

The image of flavor created in our brain raises an interesting question about the influence of cooking on the development of the human brain. Was the alluring flavor of roasted food as responsible for driving the development of bigger brains in *Homo erectus* as the increased energy and nutrients made available by cooking food? The complex processing of so much information to create the image of flavor in our mind may have played an important role. We know that comfort foods lead to more taste and smell receptors, which in turn require more processing in the brain. The receptors for sweet, umami, and bitter have been discovered fairly recently in the gastrointestinal tract of humans. Here they play a role in a variety of bodily responses such as the absorption of glucose and the release of insulin. These "taste" receptors convey messages to the brain that turn on and off various appetite hormones. It is conceivable that processing nerve responses from gastrointestinal receptors may have placed additional demands on the brain, helping to speed its development. The delicious flavor of roasted meat may also have accounted for the increased consumption of meat in the diet of early humans. It is interesting to note that humans perceive a much stronger umami taste from cooked meat than raw meat. Why is this important? Because the body can convert amino acids from the digestion of cooked meat proteins to the glucose required by the brain through *gluconeogenesis*. On average, the breakdown of 5.6–7 ounces (160–200 grams) of protein can produce the 4.2 ounces (120 grams) of glucose required for energy every day. In a healthy person the body prefers to use the amino acids from the breakdown of proteins for the synthesis of more important building blocks such as RNA, DNA, and neurotransmitters rather than the production of glucose.

Cooking makes the protein and starch in raw food much easier to digest. Heating meat over a number of hours slowly breaks down the protein collagen in tough, chewy connective tissue to easily digested, soft gelatin. Humans have a very difficult time digesting raw starch. Heating food with water converts dry starch granules to an easily digested form by a process known as gelatinization.

It is not clear when all of the digestive enzymes that break down proteins, starch, and fats, rendering them more available for absorption into the body, evolved but it was probably many millions of years ago, well before the evolution of the earliest prehuman ancestors. Digestive enzymes are complex, three-dimensional proteins that have very specific

(CONTINUED ON PAGE 14)

Starch is one of the most common forms of carbohydrates in food. All plants produce glucose from carbon dioxide and water by photosynthesis. As glucose is produced, it is converted to starch, a polymer composed of thousands of glucose molecules linked together end to end, which is then stored in plant cells for use as a source of energy. Starch is thus an efficient way of storing lots of glucose in a minimal amount of space. Mammals also store glucose as a giant polymer called *glycogen*, which has a structure that is similar to that of starch. But starch and glycogen differ in one important aspect. Plants store glucose in two forms: one is *amylose*, a smaller, linear molecule shaped like a long chain of paper clips, and the other is *amylopectin*, a much larger, branched molecule shaped like a tree with both short and long branches attached to a trunk. It is most common for plants to produce amylose and amylopectin in a weight ratio of about 1:4. However, some plants produce starch that contains almost no amylose; this is commonly referred to as *waxy* starch (see the box in chapter 3 titled The Difference Between Waxy and Mealy Potatoes). For whatever reason, mammals evolved to produce only the highly branched polymer of glucose and none of the smaller, linear form.

The amylose and amylopectin molecules produced by plants are organized into microscopic particles called *starch granules* and stored within plant cells until needed for energy. The amylopectin molecules are arranged in alternating layers of organized crystalline structures and amorphous noncrystalline structures, with the amylose molecules randomly dispersed throughout. The linear molecules of amylose and the ends of the long branches of amylopectin form helical structures, both alone and entwined together. When the helical structures pack together, they create the ordered crystalline regions within the granules. The sizes and shapes of the granules vary with each plant but are consistent within each type of plant. Figure 1 shows a micrograph of starch granules within empty potato cells. Potatoes contain the largest starch granules of the commonly consumed vegetables. They also contain a large number of granules per cell.

When starch granules are heated in water, they begin to absorb some of the water and swell, as when air is blown into a balloon. The granules continue to absorb water as the temperature rises until they reach their maximum volume and viscosity, a point called the *gelatinization temperature*. This temperature is quite specific for each type of starch, whether it is in corn, wheat, potato, rice, or sorghum. The gelatinization temperature depends on

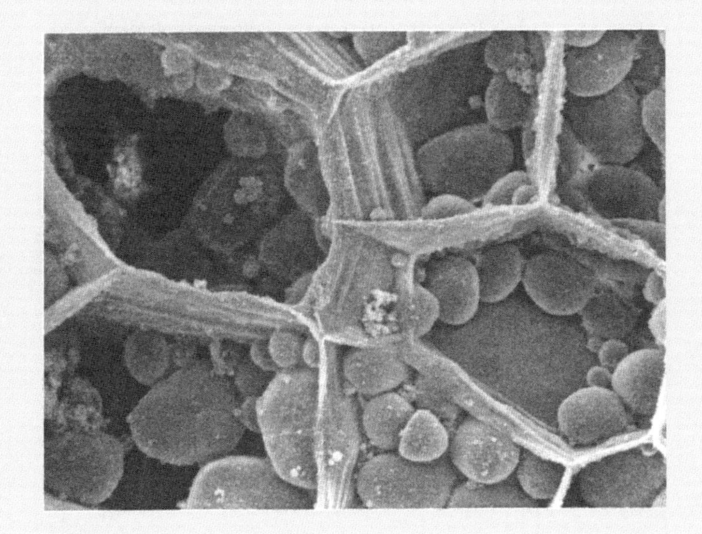

FIGURE 1

Scanning electron micrograph (SEM) of potato cells, showing large granules of starch within the cells, bounded by cell walls. Other components of the cells such as fluid have been removed. Adapted from SEM by Don Galler, Massachusetts Institute of Technology, courtesy of News.MIT.edu, August 14, 2012.

the type of starch and the ratio of amylose to amylopectin in the starch granules. Higher amylose content delays swelling and increases the gelatinization temperature. A good example is the starch in rice. The starches in short-, medium-, and long-grain rice vary in their ratios of amylose to amylopectin. Long-grain rice contains 22–28 percent amylose by weight, medium-grain rice contains 16–18 percent amylose by weight, and short-grain rice contains less than 15 percent amylose by weight and may contain almost no amylose (making it waxy starch). Varieties of long-grain rice have a gelatinization temperature above 158°F (70°C), while waxy short-grain rice gelatinizes at 144°F (62°C). The gelatinization temperature of the starch greatly affects the texture of the cooked rice. This largely explains why long-grain rice is fluffy and short-grain rice is sticky. Granules in short-grain rice burst at a much lower temperature, releasing starch molecules that cause the rice grains to stick together, while the granules in long-grain rice tend to remain intact.

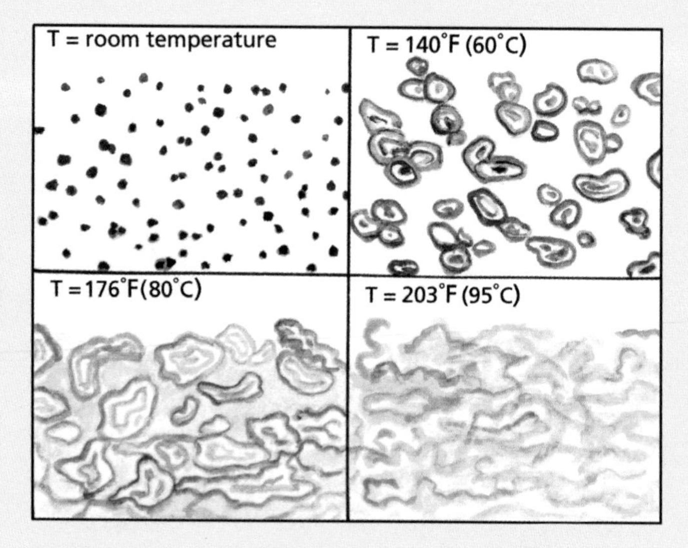

FIGURE 2

The structure of corn starch granules changes during gelatinization as they are heated in water from room temperature (uncooked starch, upper left) to 203°F (95°C) (lower right), at which point nearly all of the granules have been disrupted and have become dispersed within the water. Peak viscosity (thickness) of the gelatinized starch is achieved at 176°F (80°C), while heating to 203°F (95°C) results in significant loss of viscosity. Watercolor on paper by the author.

Figure 2 shows how granules of pure cornstarch absorb water and swell when heated. At 140°F (60°C), the granules have started to swell, and at 203°F (95°C), they have swollen so much that they are difficult to see. Eventually, they burst and release amylose and amylopectin, creating a network of entwined molecules that trap water and thicken to a gel on cooling. This is how cornstarch thickens gravies and sauces and turns them to solid gels when refrigerated.

The behavior of the starch molecules in bread causes it to stale and turn firm and dry. Fresh-baked bread contains about 35 percent water by weight, with the wheat starch granules swollen and gelatinized. At this level, the large starch and protein molecules are hydrated and flexible and are able to move about to a certain degree, making the bread soft and tender. But after a few days, the bread becomes firm and appears to dry out. Most people assume it is because much of the moisture has evaporated, but in fact it really hasn't. As bread ages, the amylose molecules form helices, which pack together like pencils in a box, forming crystalline structures that trap water molecules inside the crystalline regions. This makes the bread appear to be dry, but in fact it has not lost any significant amount of water. The crystalline structures also make the bread feel firm. The ends of the long branches of amylopectin behave in a similar way, but these crystalline regions are relatively short and not as strong as those formed by longer chains of amylose, and they will reverse with mild heat. If stale, dry, firm bread is briefly heated in the microwave, these less stable crystalline regions of amylopectin are disrupted, releasing the trapped molecules of water and making the bread appear softer and more moist, at least temporarily. The process by which gelatinized starch molecules form crystalline regions over time is called *retrogradation*. Starch molecules will retrograde (crystallize) at room temperature, but they undergo this process at a much faster rate in the refrigerator. So don't place bread in the refrigerator to keep it from staling; this only hastens the process. Fortunately, refrigerated bread can be revived by briefly microwaving it. Bread can also be stored in the freezer. Once the water molecules are frozen, the starch molecules are also frozen in space, curtailing retrogradation. Frozen bread can be thawed to yield soft, moist bread.

One final point about starch in food. Most starch, especially if it has been heated and gelatinized, is rapidly digested to glucose, which is quickly absorbed into the body, elevating the levels of glucose and insulin in the blood. The amount of glucose absorbed over a period of several hours following consumption of a food is called the *glycemic index* of that food. Foods with a higher glycemic index cause the rapid release of more insulin into the blood system, which impacts the amount of fat stored in fat cells. Retrograded starch is very poorly digested by digestive enzymes, so much of it passes into the large intestine, where it is digested by the gut bacteria, which convert it to *short-chain fatty acids*, such as butyric acid and propionic acid. The cells lining the large intestine use these short-chain fatty acids for energy. Retrograded starch, also called *resistant starch* because it is resistant to digestion, functions as a prebiotic and is beneficial for the colonic cells. Because retrograded starch is not digested to glucose, the calorie content of foods containing retrograded or resistant starch is lower than that of cooked high-starch foods like rice, potatoes, and fresh white bread. The amount of retrograded starch is directly related to the amylose content of the starch. Legumes are especially high in amylose, and therefore resistant starch, which is a major reason why they are a healthy food. Eat beans for resistant starch!

REFERENCES

Buleon, A., P. Colonna, V. Planchot, and S. Ball. "Starch Granules: Structure and Biosynthesis." *International Journal of Biological Macromolecules* 23 (1998): 85–112.

Crosby, G. A. "Resistant Starch Makes Better Carbs." *Functional Foods and Nutraceuticals* (June 2003): 34–36.

Tester, R. F., and W. R. Morrison. "Swelling and Gelatinization of Cereal Starches. I. Effects of Amylopectin, Amylose, and Lipids." *Cereal Chemistry* 67, no. 6 (1990): 551–557.

Thomas, D. J., and W. A. Atwell. *Starches*. St. Paul, MN: Eagan Press, 1997.

The human immune system evolved as protection for the body against the invasion of large foreign molecules such as proteins and complex polysaccharides. About 95–98 percent of food by weight is composed of the four macronutrients: water, carbohydrates, proteins, and fats. Except for water and simple sugars, the macronutrients are all very large molecules that cannot be absorbed into the body without first being broken down into much smaller molecules by digestive enzymes in the small intestine (most of which are formed in the pancreas). These molecules can then be transported across the cell membranes lining the gastrointestinal system. The food we eat is therefore not available to the body until it is first digested to smaller molecules in the gastrointestinal system. Complex polysaccharides such as starch must be broken down to simple sugars, proteins converted to amino acids and small peptides (two or three amino acids linked together), and fats separated into individual fatty acids and glycerol.

Digestion of food starts in the mouth, where some of the starch is digested to simple sugars. The food then moves to the stomach, where very little absorption of nutrients occurs, and next to the small intestine, where most digestion takes place. Simple water-soluble molecules such as sugars and amino acids are transported directly into the portal blood system and then through the liver (where many are metabolized), while fat-soluble fatty acids and other lipids such as cholesterol and fat-soluble vitamins are first transported into the lymph system before entering the blood system just before it passes into the heart. The small intestine consists of three sections: the duodenum, jejunum, and ileum, in that order from top to bottom. The total length of the small intestine is about 10 feet, with a surface area that is greatly enlarged due to the presence of numerous folds and tiny fingers (villi); they provide a surface area for the absorption of nutrients that is about six hundred times larger than the surface of a cylinder of the same diameter and length as the small intestine. It takes about 4 hours for an undigested portion of a meal to pass from the mouth through the small intestine and enter the large intestine, with all of it entering within about 8 hours of consumption.

What isn't digested and absorbed in the small intestine passes into the large intestine (colon) where it is metabolized (fermented) by billions of mostly beneficial bacteria. It enters the ascending portion of the large intestine from the ileum, moves on to the transverse section, and finally reaches the transcending portion, where it is excreted as feces. Transport and metabolism of food residues through the large intestine are much slower than through the small intestine, often taking up to a week and more to be excreted. Most fermentation of food residues by bacteria occurs in the ascending colon, while water and water-soluble salts are absorbed from the descending colon, where the feces are also formed, with as much as 60 percent of the weight composed of dead bacterial cells.

Food residues entering the large intestine are primarily composed of nondigestible dietary fiber (mostly complex polysaccharides), which includes about 1–10 percent of the starch that is ingested with the food that is resistant to digestion in the small intestine. So-called resistant starch is prevalent in whole cereal grains and legumes, and it either is protected from digestion by association with proteins or possesses a crystalline structure that makes it inaccessible to the digestive enzymes in the small intestine. Grinding cereal grains such as whole wheat into refined flour breaks up the protein-starch complexes and makes much more of the starch available for digestion in the small intestine. Cooking cereal grains and legumes breaks down most of the crystalline structure of starch, making it much more susceptible to digestion. Nearly all resistant starch that enters the large intestine is metabolized by the gut bacteria to short-chain fatty acids (with two to four carbon atoms) that supply about 60 percent of the energy required by the cells lining the lower colon. During this process, the bacteria produce hydrogen, methane, and carbon dioxide gases from the soluble nondigested dietary fiber and resistant starch, which can cause unpleasant gas and bloating. All humans have a difficult time digesting raw starch; however, cooking high-starch foods such as tubers made more nutrients available for energy, as well as making them more tolerable for consumption by early humans.

The Recommended Daily Allowance for protein is about 56 grams per day for a 150-pound (70-kilogram) modern adult human. Today's typical Western diet provides about twice the required amount of protein, all of which is efficiently digested and absorbed. It is difficult to say how effective very early humans like *Homo erectus* were at digesting protein, but more than likely all the enzymes for efficiently digesting protein (called protease enzymes) were in place by the time of their evolution nearly 2 million years ago. The presence of small proteins called *trypsin inhibitors* in peas, beans, lentils, wheat, buckwheat, and rice bran may have affected the ability of very early humans to completely digest proteins and may have lowered the nutritional quality of the proteins they consumed. Fortunately, cooking makes proteins easier to digest, and trypsin inhibitors are rendered inactive by cooking these foods in hot water. Thus it becomes relevant to know when food was first cooked by boiling in water. This is another example of how cooking may have influenced the evolution of early humans.

REFERENCE

Tso, P., and K. Crissinger. "Overview of Digestion and Absorption." In *Biochemical and Physiological Aspects of Human Nutrition*, ed. M. H. Stipanuk, 75–90. Philadelphia: Saunders, 2000.

requirements for the structure of the proteins, polysaccharides (such as starch), and fats that they degrade into smaller fragments for absorption into the body. Proteins are degraded to smaller peptides and amino acids by protease enzymes, starch is degraded to the simple sugars maltose and glucose by amylase enzymes, and fats are degraded to free fatty acids by lipase enzymes. The structures of these enzymes almost certainly morphed into more active forms as humans evolved. Our earliest ancestors, who lived 4–5 million years ago and earlier, ate mostly plants (we know this from isotopic analysis of teeth enamel) and probably evolved the enzymes to digest starch before developing the enzymes to efficiently digest proteins and fat. The very early evolution of enzymes to digest plant-based starch to glucose may have played a role in the brain's use of glucose as its primary source of energy. As early humans consumed more meat, the enzymes for digesting protein and fat changed structurally to become more efficient, thus allowing proteins and fats to contribute more energy and nutrients for the body and leading to the development of bigger brains.

Early Methods of Cooking Food

Clearly, the controlled use of fire to cook food was an extremely important element in the biological and social evolution of early humans, whether it started 400,000 or 2 million years ago. The lack of physical evidence suggests early humans did little to modify the control and use of fire for cooking for hundreds of thousands of years, which is quite surprising, given that they developed fairly elaborate tools for hunting during this time, as well as creating some of the first examples of cave art about 64,000 years ago. Physical evidence shows that cooking food on hot stones may have been the only adaptation during the earliest phases of cooking. Then, about 30,000 years ago, "earth ovens" were developed in central Europe. These were large pits dug in the ground and lined with stones. The pits were filled with hot coals and ashes to heat the stones; food, presumably wrapped in leaves, was placed on top of the ashes; everything was covered with earth; and the food was allowed to roast very slowly. The bones of many types of animals, including large mammoths, have been found in and around ancient earth ovens. This was clearly an improvement over rapidly roasting meat by fire, as slow cooking gives time for the collagen in tough connective tissue to break down to gelatin; this process takes at least several hours, and often much longer, depending on the age of the animal and where the meat comes from in the animal. The shoulders and hindquarters of animals are involved in more muscular action and thus contain more connective tissue than the tenderloin near the ribs. Breaking down

FIGURE 1.2

Ancient Chinese earthenware tripod cooking vessel, twelfth century B.C.E., from the collection of the Metropolitan Museum of Art, New York. Note that the combed ridges increased heating efficiency by increasing the vessel's surface area.

tough connective tissue makes the meat easier to chew and digest. Like today's barbeque methods, cooking meat slowly in earth ovens made it very tender and flavorful.

After dry roasting with fire and heating on hot stones, the next true advance in very early cooking technology appears to have been the development of wet cooking, in which food is boiled in water. Boiling food would certainly be an advantage when cooking starchy root tubers and rendering fat from meat. Many archeologists believe the smaller earth ovens lined with hot stones were used to boil water in the pit for cooking meat or root vegetables as early as 30,000 years ago (during the Upper Paleolithic period). Others believe it is likely that water was first boiled for cooking in perishable containers, either over the fire or directly on hot ashes or stones, well before this time. Unfortunately, no direct archeological evidence has survived to support this conclusion. Yet we know that even a flammable container can be heated above an open flame as long as there is liquid in the container to remove the heat as the liquid evaporates. Thus containers made of bark or wood or animal hides could have been used for boiling food well before the Upper Paleolithic period. No physical evidence of sophisticated utensils for cooking food appears until about 20,000 years ago, when the first pieces of fired clay pottery appear. Using sensitive chemical methods, scientists have determined that shards of pottery found in Japan contain fatty acids from marine sources such as fish and shellfish. These heat-resistant pots may have been used to boil seafood. The development of simple clay ovens did not occur until at least 10,000 years later. If cooking has had such a profound effect on the evolution of humans, why is there little evidence from earlier periods of the development of more sophisticated methods of cooking than simply roasting in a hot pit or boiling in water with hot stones?

Jacob Bronowski may have answered that question in his enlightening book *The Ascent of Man*. The life of early nomads, such as the hunter-gathers who existed for several million years or more, was a constant search for food. They were always on the move, following the wild herds. "Every night is the end of a day like the last, and every morning will be the beginning of a journey like the day before," he wrote. It was a matter of survival. There simply was no time for them to innovate and create new methods of cooking. Being constantly on the move, they couldn't pack up and carry heavy cooking utensils every day, even if they had invented them. Then, about 10,000 years before the last ice age ended, creativity and innovation finally began to flourish in spite of the restrictions of nomadic life. Early humans were finding that food was becoming more abundant due to warming weather, so they could gather it more easily without needing to move constantly.

2 The Dawn of Agriculture Revolutionizes Cooking (12,000 Years Ago–1499)

The Dawn of Agriculture Changes Everything

With the end of the last ice age and the beginning of the Neolithic period, about 12,000 years ago, everything changed. *Everything!* It was the dawn of the agricultural revolution, when wandering nomads began to settle and turn into villagers. What made this possible? The discovery that seeds from new varieties of wild grasses that emerged after the end of the ice age, such as emmer wheat and two-row barley, could be gathered, saved, planted, and harvested the following season. This occurred first in an area known as the Fertile Crescent (Jordan, Syria, Lebanon, Iraq, Israel, and part of Iran). Enough food could now be harvested in 3 weeks to last an entire year! Being able to harvest large quantities of food at one time meant these early farmers could no longer move from place to place; they had to build immovable structures for storing and protecting all the food, and this resulted in the creation of permanent settlements. The agricultural revolution then spread to other parts of the world over several thousand years. Thanks to the pioneering research of the Russian scientist Nikolai Vavilov in the 1930s and the American scientist Robert Braidwood in the 1940s, we now know that over several thousand years people living in seven independent regions of the world domesticated crops and animals indigenous to that region (for a summary, see table 2.1). Unfortunately, Vavilov's studies were prematurely ended when he was imprisoned in 1940 by the Stalinist government for his revolutionary views on evolution.

As the ice age was coming to an end around 12,000 years ago, early humans were harvesting wild wheat and barley in quantity in the Fertile Crescent, but there was no evidence of

FIGURE 2.1

The Harvesters, by Pieter Bruegel the Elder, 1565, from the collection of the Metropolitan Museum of Art, New York.

TABLE 2.1
The Early Domestication of Plants and Animals by Region and Time

Independent Region	Domesticated Plants and Animals	Relative Time Frame (Years Ago)
Fertile Crescent	Wheat, barley, goats, sheep, cattle	10,000–8,000
Southern China	Rice, pigs	8,500
Northern China	Chickens, millet	7,800
Central America	Corn, beans	7,500
South America	Potatoes	5,000
Sub-Saharan Africa	Sorghum	5,000
North America	Sunflower	4,500

Source: Information compiled from B. D. Smith, *The Emergence of Agriculture* (New York: Scientific American Library, 1995).

domesticated plants and animals. By domesticated, I mean plants and animals deliberately raised for food by humans rather than wild plants and animals gathered in the forests and fields. Then within a period of roughly 300 years, between 10,000 and 9,700 years ago, the first evidence of domesticated plants and animals began to appear in the southern Jordan Valley around the ancient settlement of Jericho. In this relatively brief time period, the seeds of plants like wheat and barley became larger while the bones of animals became smaller. That's how archeologists in the field can tell the difference—and it makes sense. As early humans began to select seeds to plant, they chose the larger seeds, which were storing more of the nutrients required for faster growth. The resulting crops grew faster to outcompete the wild weeds and provided higher yields—and in turn produced still larger seeds. These early humans also selected wheat plants with terminal clusters of seeds that retained the kernels during harvest instead of allowing them to scatter in the wind like the wild varieties. The rachis, the short stalk that holds the seed to the plant, became shorter and thicker with time (figure 2.2). DNA analysis confirms that the physical differences observed between domesticated and wild seeds originate in the plant's genome. All these changes occurred as a result of human selection of plants with more desirable traits. These are the first plants to be genetically modified through human intervention. Similarly, domesticated goats and sheep were selected to be more docile and adaptable to living in a confined pen and feeding off the scraps of food left by their keepers. Thus they became smaller. These physical changes in domesticated plants and animals began to take shape as humans started to produce their own food.

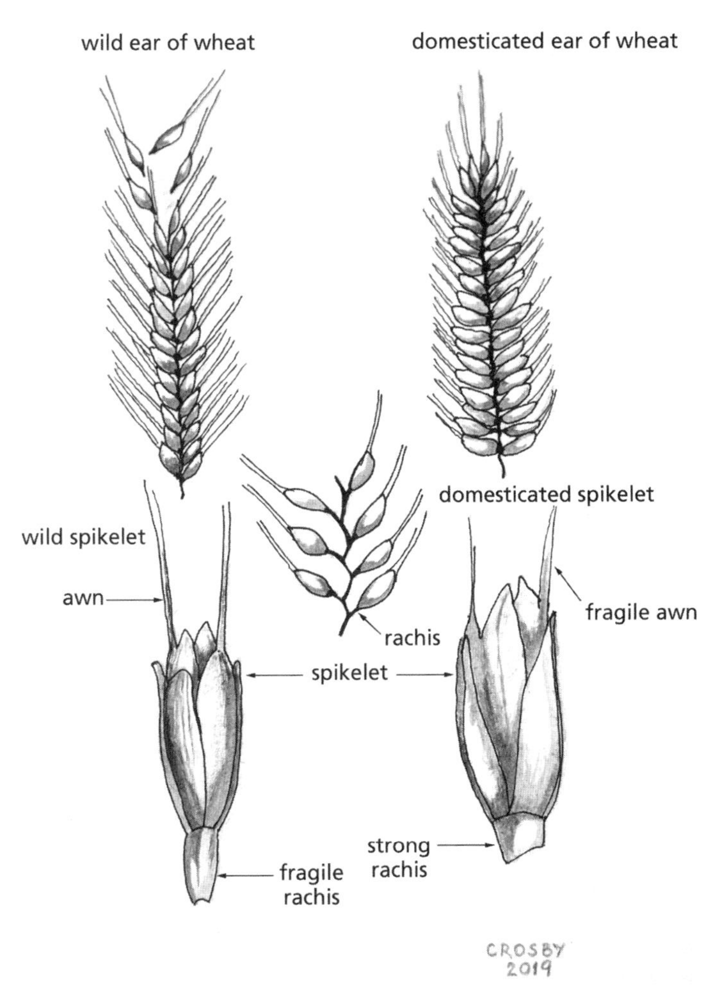

FIGURE 2.2

A comparison of wild and domesticated wheat. Note the difference in the sizes of the rachises that hold the seed (spikelets) to the plant. The larger rachis of the domesticated wheat reduces the tendency of the seed to scatter in the wind. Ink and watercolor on paper by the author, based on an illustration in *The Emergence of Agriculture* by Bruce D. Smith (Scientific American Library, 1995), 73.

The development of new foods and methods of cooking in the few thousand years following the emergence of agriculture illustrates how important this period was for the advancement of humans. The change from a nomadic life to a sedentary life in more secure settlements was critical, as it allowed humans to make significant achievements in technology and other areas. Within a few thousand years, small farming villages grew into large permanent settlements and then small cities. Jericho is perhaps the oldest permanent settlement, providing an accurate record of agricultural development between 10,000 and 9,700 years ago. Hunter-gatherers first settled there around 11,000 years ago in order to be near a constant source of water, a spring-fed oasis. Archeological excavations of the oldest buried sections of Jericho, which cover an area of a little less than ¼ acre (0.1 hectares), did not reveal any signs of domesticated seeds or animal bones. By 9,700 years ago, the first domesticated seeds of emmer wheat and barley began to appear in higher levels of soil, and the earliest farming settlement had grown to an area of about 6 acres (2.5 hectares) with perhaps 300 people living in mud brick houses. By 8,000 years ago, Jericho was home to a permanent agricultural settlement of approximately 3,000 people occupying an area of 8–10 acres (3.2–4 hectares). About this same time, emmer wheat hybridized with a wild grass to produce bread wheat, which contained higher levels of the gluten-forming proteins required for making leavened bread. Wheat had finally emerged in the form in which it is still grown and used today around much of the world.

EXPLAINING GLUTEN

One question I hear a lot is "What is gluten? Can you please explain it to me?"

It's not surprising that people have a difficult time understanding gluten because it doesn't exist in nature. Gluten is a water-insoluble *protein* that is formed when water is mixed with wheat flour. Proteins are very large molecules composed of amino acids. Two of the naturally occurring proteins in flour are called *glutenin* and *gliadin*. When sufficient water is added to dry flour, the two proteins emerge from a "frozen state" and become flexible and able to move about. It's not too different from stiff, dry spaghetti becoming flexible when it is cooked in water.

The process of wetting the proteins is called *hydration*. As water and flour are mixed, the hydrated proteins are brought together and begin to interact. They literally stick to each other through the formation of chemical bonds. These new chemical bonds are called *cross-links*. Again, it's not unlike cooked spaghetti after it has been poured into a colander and allowed to drain. After a few minutes, the strands of spaghetti stick together unless they are separated by the addition of a little oil. In the case of gluten, a number of different types of chemical bonds form between the proteins, with some (disulfide bonds) being stronger than others (ionic bonds and hydrogen bonds).

Continued mixing causes more cross-links to form between the proteins until a large *network* of chemically linked proteins is formed. Mixing can be done with a stand mixer or by hand, as when dough is *kneaded*. When dough is mixed or kneaded, the hydrated flexible proteins are stretched and aligned in the direction of kneading, providing more opportunities to form cross-links between the proteins. Kneading also incorporates air, which helps to form strong disulfide bonds. As kneading continues, the protein networks combine to form *sheets* of proteins. Think of this step in the process as being similar to unraveling thread (proteins) and then weaving the straightened thread into pieces of cloth (networks) and finally stitching the pieces of cloth together to form large sheets of cloth similar to a quilt (see figure 1).

The chemical cross-linking of glutenin and gliadin forms gluten, a very *elastic* substance. Neither protein alone is as elastic and stretchable as gluten. But when they are chemically linked together, the resulting new protein becomes elastic and stretchable like a rubber balloon. The gluten becomes stronger and stronger as more bonds form between the proteins, like a weak, thin rubber balloon becoming a strong, thick rubber balloon. And like a balloon, gluten can be inflated with gas and steam as dough rises and bread is baked in the oven. Think of an inflated balloon covered with papier-mâché, which is how we made puppets when I was young. When the papier-mâché dries, the balloon can be popped and removed, leaving a rigid sphere. In the same way, as bread bakes, the inflated gluten dries, turning into a strong but flexible structure that creates the holes in chewy bread. (Note that freshly baked bread still contains about 35 percent moisture by weight, which is sufficient to keep the starch soft and the gluten elastic.)

The *scanning electron micrographs* (SEMs) of different stages of gluten development in figure 1 were taken under high magnification and may help you envision the formation of gluten from flour and water. The dimension bar at the bottom of each SEM shows distance in microns (a micron is one-millionth of a meter).

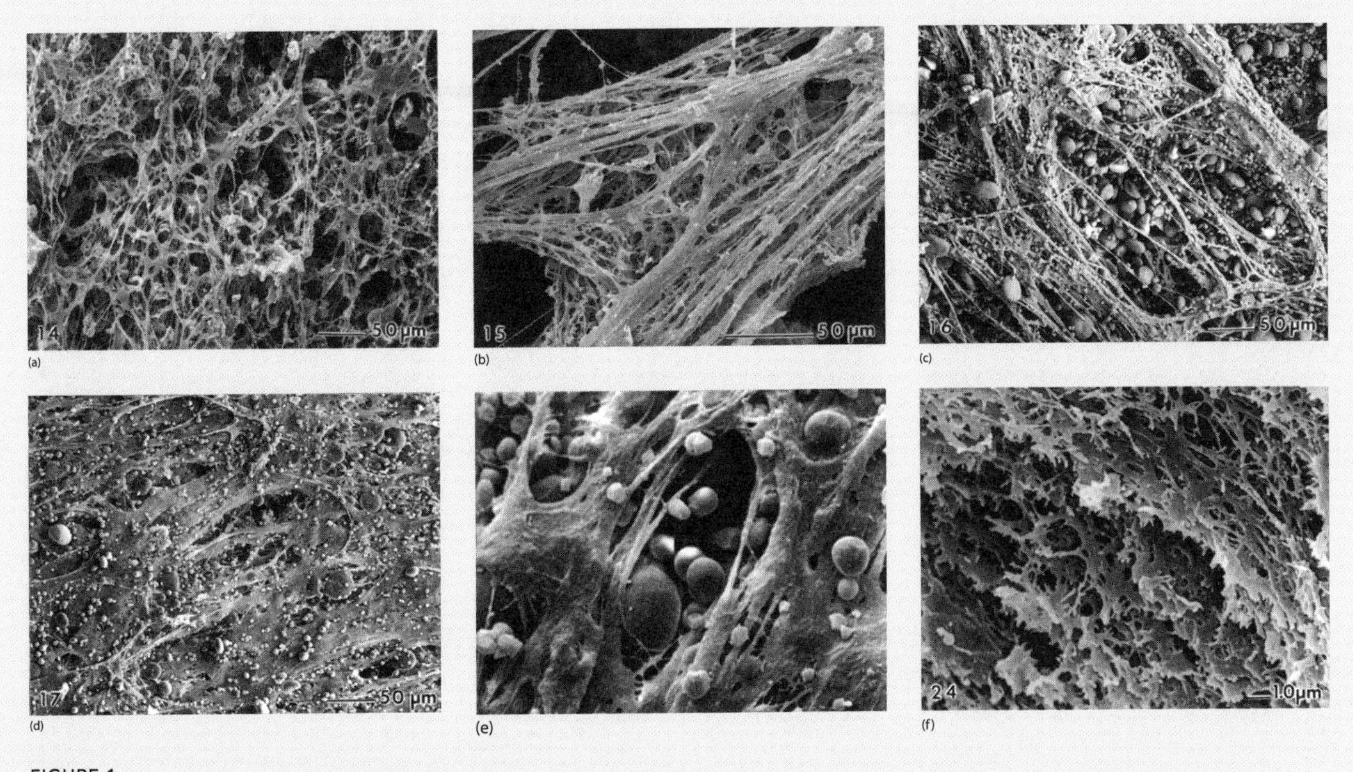

FIGURE 1

(a) A scanning electron micrograph (SEM) shows the protein network formed after water is added to wheat flour with no mixing. Starch granules have been removed by washing (the starch granules normally fill the holes). (b) The protein strands begin to stretch and bond together after a few seconds of kneading the dough. Starch granules have been removed by washing. (c) Underkneaded dough is shown with stretched aggregated strands of protein. Starch granules have been washed away at the surface but remain trapped within the protein matrix. (d) A protein network results from optimally kneaded dough. Starch granules have been washed away at the surface. (e) Another SEM shows a higher magnification of optimally kneaded dough with the starch granules still in place. When the dough bakes, the starch granules will absorb water and swell many times larger, filling most of the gaps. (f) Under higher magnification, an exploded view shows how sheets of gluten layer upon each other to form a flexible membrane capable of holding in gas. Photomicrographs reproduced from *Zeitschrift fur Lebensmittel-Untersuchung und -Forschung* 190 (1990): 401–409.

Many factors affect the development and strength of gluten. If too much gluten develops, it can make baked goods, like piecrust, tough rather than tender. The extent of gluten development also affects how high leavened baked goods will rise and whether the crumb will be tender or chewy. Weaker gluten is more extensible (stretchable) and doesn't shrink as much. Here are some factors that affect gluten development:

1. Variety of wheat: Soft wheat contains less protein (6–8 percent), less glutenin, and smaller proteins and forms weaker gluten. Hard wheat contains more protein

(10–14 percent), more glutenin, and larger proteins and forms stronger, more cohesive, elastic gluten.

2. Amount of water: Hydration is essential for gluten development. Glutenin and gliadin absorb about twice their weight in water (hydration). Less water results in less gluten development by reducing protein mobility, but too much water also reduces gluten development by diluting the proteins so much that their interaction is restricted.
3. Water hardness: The calcium and magnesium in hard water strengthen gluten. Water in Boston is soft, containing zero to sixty parts per million of calcium and magnesium.
4. Water pH: The ideal pH for gluten development is 5–6. A pH above or below pH 5–6 reduces gluten strength, producing more extensible (stretchable) dough. Adding baking soda raises pH, producing more cookie spread and a more porous, tender crumb.
5. Leavening: Expanding air bubbles strengthen gluten, increasing cohesiveness and elasticity and producing higher volume and a finer crumb.
6. Enzymes: Enzymes that break down proteins are naturally present in flour but are inactive when dry and only become active when hydrated with water. Enzymes break down gluten into smaller pieces, so dough becomes softer and more extensible. Resting dough for 15–30 minutes allows time for this process, known as *autolysis*, during which the enzymes break down gluten to produce more extensible dough, providing more volume and an open crumb.
7. Salt: Bread dough contains 1.5–2.0 percent salt by weight of flour. Salt slows enzyme activity and rate of fermentation. It also strengthens gluten, producing bread with higher volume and a finer crumb.
8. Fat, oil, emulsifiers, and sugars: Fat, oil, emulsifiers, and sugar tenderize dough. Fat and emulsifiers coat proteins, reducing hydration and gluten development (like oil coating spaghetti). Shortening, like lard or vegetable oil, literally shortens gluten strands (which is the origin of the word "shortening"), producing more tender baked goods. Sugar competes for water, reducing protein hydration and gluten development.

REFERENCE

Amend, T., and H.-D. Belitz. "The Formation of Dough and Gluten—A Study by Scanning Electron Microscopy." *Zeitschrift fur Lebensmittel-Untersuchung und-Forschung* 190 (1990): 401–409.

Similar developments were happening in other parts of the world. Evidence of rice agriculture dating from 7,800 to 8,400 years ago was found at Pengtoushan in the Hubei basin near the origin on the Yangtze River in China. In this same village, the earliest house structures discovered in China have been dated by radiocarbon analysis at 7,815–8,455 years ago, a sure sign that permanent settlements were occurring in other parts of the world around the same time as Jericho. Near P'ei-li-kang, in the Yellow River region of northwestern China, millet was first cultivated sometime between 9,000 and 7,500 years ago when inhabitants of this area transitioned from hunter-gatherers to farmers. The starch-rich millet was ground into flour with footed stone mortars and pestles and used to prepare many types of foods. The most interesting for our purposes is the earliest form of noodles found recently in an earthenware bowl at Lajia in Qinghai, China, and dated by radiocarbon analysis to be 4,000 years old. It's as though someone's meal of freshly made noodles had suddenly been interrupted by a natural disaster such as a flood or earthquake. These long, thin noodles predate by 2,000 years the noodles made from more traditional wheat found in Italy.

The Greatest Technological Advance of All Time

Many anthropologists consider the deliberate production of food to be the greatest technological advance of all time. Why? Because in the 5,000 to 7,000 years following the emergence of agriculture, the world's population exploded from an estimated 3 million people to 100 million people! In this relatively brief period, the population of early humans grew almost thirty-five times more than it had during the previous 4–5 million years. Permanent settlements offered safety and more time for childbirth. They made it possible to domesticate animals and other crops, ensuring a steady supply of meat and dairy products that provided more energy and better nutrition and allowed more time for innovation, including the creation of new foods and methods of cooking. The development of agriculture fostered innovation, and from this, the first scientific principles evolved for acquiring new knowledge based on observable facts.

This is the time when we first see aboveground clay ovens created to bake bread from dough made of ground kernels of bread wheat and water. It is conceivable that seeds from wild grasses were ground, mixed with water, and baked on hot stones long before the use of domesticated wheat. Either way, learning to make bread is an amazing discovery that may have come about by much trial and error, signaling the first scientific approach to creating a new form of food by cooking, tasting, and learning how to improve the earliest forms of bread. It seems miraculous that wheat was one of the first wild grasses to be domesticated because it is the only cereal grain containing enough of the gluten-forming proteins required for making leavened bread. The proteins in other cereals are not as effective. Making bread from wheat also required the development of heavy stone implements for grinding hard wheat kernels into finely grained flour. It meant that wheat and other cereal grains became a substantial part of the diet, constricting that diet to a narrower range of domesticated foods compared with the preagricultural period. The long-term implications of the increased consumption of cereal grains as a substantial component of the human diet are still being studied and debated.

The ancient Egyptians were the first to use yeast to make leavened bread, beginning as early as 5,000 years ago. They adapted the technology from their use of yeast to make alcoholic beverages, a process they discovered about the same time. Borrowing from the Egyptians, the ancient Greeks are credited with significantly improving the methods for making flour, bread dough, and bread. They produced white flour with the use of sieves to remove the bran, which reduces the ability of yeast-leavened bread to rise. They also developed methods to grind bread wheat into flour to make semolina bread. It is reported the Greeks developed about seventy different types of bread, making it one of the most consumed foods of that time. The first front-loaded clay ovens for baking bread were also developed in ancient Greece. They were not very different from the wood-fired pizza ovens used today. Not to be outdone, the Egyptians took this one step further, creating ovens made of mud bricks sometime around 4,500 to 3,800 years ago. Rather than being dome shaped like front-loaded clay ovens, many of these ovens were made with flat tops. The flat tops evolved into cooking surfaces for heating pots and pans, and with that advance, the "cooking range" was created, followed about 3,000 years ago by what is known as a brazier. A brazier is much smaller than a brick oven, is heated with burning coals, is portable like a hibachi, and may have been used for both heating and cooking. The earliest examples were made of clay, but these were soon replaced by ones made of more rugged bronze at the beginning of the Bronze Age.

The Emergence of Agriculture Catalyzes Advances in Cooking

It is believed the ancient Egyptians made the next big advance in cooking, using palm oil to fry food, at least 3,000 years ago when they learned how to extract the oil from the fruit of the oil palm tree. Thus, cooking evolved from dry roasting, to boiling in water, to frying in fats and oils. Early humans may have been cooking with animal fat much earlier, but this is not certain. Boiling water maintains a constant temperature of 212°F (100°C) at sea level. Even steam remains a constant 212°F although it contains about five times more heat energy than boiling water.

The boiling point of water limits the highest temperature at which food can be cooked to 212°F. Solid animal fats and liquid oils can reach much higher temperatures, around 338°F (170°C), enabling new forms of cooking and new flavors. The chemistry of flavor development in boiling water is quite different from that in nonaqueous frying oils. That's why boiled beef tastes very different from beef sautéed in oil and boiled potatoes taste very different from French fried potatoes. Frying oils undergo oxidation, producing unique flavors absorbed by the food as the oil displaces some of the water in the food. Each method of cooking, dry roasting with fire, boiling in water, and now frying in oil, produces a distinctly different flavor. Records show that not long after the Egyptians used palm oil for cooking, the Greeks began using olive oil for deep-frying food. It is believed the olive tree was first domesticated about 5,000 years ago, and the technology for extracting oil from olives soon followed. Competition and technology exchange between ancient Greece and Egypt proved to be very beneficial for the early advancement of cooking science.

But frying food in oil brings a new challenge: what utensil to cook with. As noted earlier, boiling water is less of a challenge because the evaporating water takes away heat; this means almost any type of vessel can be used for cooking food in boiling water. But frying oils do not evaporate when heated, so they do not remove heat from the vessel. A fireproof vessel capable of being heated on the flat surface of a brick oven or a brazier is required. Coarse clay pots had been developed between 7,000 and 8,000 years ago, but the earliest examples were probably used to store liquids rather than cook foods. However, we have found evidence in Greece of ceramic pots and pans made roughly 3,000 years ago using finer clays fired at higher temperatures, so they were more suitable for frying as well as boiling. Recall that early humans living on the islands of Japan had created fired clay pottery more than 15,000 years earlier. It may have taken this long for the technology to find its way to the Mediterranean region. More likely though the technology for making fireproof vessels evolved from the technology used for making ovens with clay in the relatively nearby Fertile Crescent.

WHAT IS SO SPECIAL ABOUT WATER IN FOOD?

Water plays many very important roles in food. It affects texture (ranging from dry and brittle to moist and soft), enables enzyme activity and chemical reactions (acting as a solvent), supports the growth of microorganisms, makes it possible for large rigid molecules like polysaccharides and proteins to become flexible, move about and interact, and conducts heat within food.

Foods such as meat, poultry, seafood, fruits, and vegetables contain very high levels of water (75 percent or more), so water is the most abundant component in many fresh foods. Other foods, such as dairy products and fresh baked goods, also contain high levels of water (about 35 percent or more). Foods that are high in moisture are at risk of contamination from the growth of microorganisms such as bacteria, yeast, and mold, while dry foods like flour and pasta generally have long shelf lives.

But not all water that occurs in food is alike. Most food scientists divide water into three forms: *free*, *adsorbed*, and *bound*. Free water is the water that can literally be squeezed out of a food, like the juice in an orange or the water that sometimes separates in sour cream or yogurt. Adsorbed water (spelled with a *d*) is water that is attached to the surface of molecules like polysaccharides and proteins. It is not readily squeezed out of the food. When food scientists refer to the hydration of proteins such as gluten and carbohydrates such as starch, they are referring to adsorbed water. Bound water is water that is physically trapped within crystals, such as crystalline starch, or other substances in food (some food scientists prefer to define just the free and bound forms of water). The important point is that free and adsorbed water can promote the growth of microorganisms, while bound water cannot.

Food scientists measure the amount of water that is available for the growth of microorganisms, as well as for enzyme and chemical reactions, through a number known as *water activity* (a_W). Water activity is the ratio of the vapor pressure (P) of water in a food divided by the vapor pressure of pure water (P_o) at the same temperature and is measured on a dimensionless scale of 0 to 1.0. The a_W of pure water is equal to 1.0. Another way of determining water activity is by measuring the relative humidity (RH) of the atmosphere in equilibrium with the food: RH (%) = 100 × a_W. In other words, water activity is a measure of the free and adsorbed water in a food—that is, the water that is available to be converted to vapor. There is a general correlation between the moisture content of food and its water activity, as shown in the following table:

Food	Moisture %	Water Activity
Fresh meat	70	0.98
Bread	40	0.95
Flour	14	0.70
Pasta	10	0.45
Potato chips	2	0.10

Most foods will not support the growth of bacteria if their water activity is less than 0.85 because at this level there is not enough water available for the bacteria to grow. However, yeasts can grow at water activities as low as 0.70, and some molds will grow even at water activities as low as 0.60! Foods with water activities in this range usually have preservatives added to prevent the growth of yeasts and molds. Acidic foods with a pH less than 4.6, such as tomato sauce, retard the growth of microorganisms. Thus an acidic food with a water activity less than 0.85 is relatively shelf stable, especially if it is stored in the refrigerator. In this case, low pH, water activity, and temperature combine to provide good insurance against the growth of harmful pathogens.

Water activity is also related to the texture of food. The amount of moisture in a food determines the mobility of the molecules in that food, especially the large molecules like proteins and polysaccharides that need moisture to move about. Proteins and polysaccharides provide the structure of food. If they are rigid, the food will be hard, but if they are flexible, the food will be soft. In this case, water acts as a *plasticizer* (a softening agent), lowering the *glass transition temperature* (*Tg*) of the molecules in food. The *Tg* is the temperature at which molecules change from a rigid, glass-like structure (below the *Tg*) to a flexible structure (above the *Tg*). It's not too different from the melting point of a substance, such as the temperature at which a solid like chocolate melts to a liquid (in this example, the fat crystals melt rather than water). The water activity and the *Tg* of food show a steady linear relationship over a wide range of values, so it is not surprising that a food with low moisture has a low water activity and a relatively high *Tg*. A food with these characteristics is hard and crisp at room temperature. Adding more moisture, which could happen in a humid environment, increases the water activity and lowers the *Tg*, so the food becomes soft and moist at room temperature. In other words, the *Tg*, the temperature at which the molecules change from flexible to rigid, is now below room temperature. This explains why foods become rigid and hard in the freezer.

But why does a simple molecule like water (H_2O) play so many important roles in food? It all has to do with *hydrogen bonds*, which occur between water molecules and between water molecules and other molecules that contain oxygen and nitrogen atoms, such as proteins and carbohydrates. A molecule of water has a unique structure, with two hydrogen atoms bonded to a single oxygen atom. Rather than being a linear molecule (H-O-H), the two oxygen-hydrogen bonds are separated by an angle of 104.5 degrees due to the repulsion of other electrons (called nonbonded electrons) in the oxygen atom. But, more importantly, oxygen is a very *electronegative* atom. It has a strong affinity for electrons, while hydrogen does not. So the two electrons in each oxygen-hydrogen bond of a water molecule spend more time around the oxygen atom, giving the oxygen atom a partially negative electrical charge, while each hydrogen atom carries a partially positive electrical charge. This produces a strong electrostatic attraction between the hydrogen atom of one water molecule and the oxygen atom of another water molecule, creating a hydrogen bond between two molecules of water. Because there are two hydrogen atoms in each water molecule, one water molecule can actually hydrogen bond with two other water molecules and so on, thus creating an infinite network of hydrogen bonds between all the water molecules in a container of water.

The hydrogen bonds between water molecules are considered to be relatively weak, being only about 5 percent as strong as the chemical bonds formed between the oxygen and hydrogen atoms in a molecule of water. Yet when all the hydrogen bonds are considered within a "sea" of water molecules, it becomes clear why it takes so much energy to separate molecules of water from each other so they can begin to move more rapidly as heat is applied. It takes twice as much energy to raise the temperature of water by, say, 68°F (20°C) as it takes to raise the temperature of olive oil by the same number of degrees. This also explains why it takes five times more energy to physically separate water molecules to the point that they can escape from each other as steam. When water molecules escape as steam, they take all this extra energy with them, so the boiling point of water never rises above 212°F (100°C) at 1 atmosphere of pressure. If more heat is added to boiling water, the water just boils and turns to steam more quickly rather than becoming hotter. But when water freezes, there is no place to which the molecules can escape, so ice can be cooled to virtually any temperature below 32°F (0°C). Ice cubes in a freezer will be the same temperature as the freezer.

Let's finish our discussion of the roles of water in food by examining how water influences the properties of proteins and polysaccharides. As mentioned earlier, foods with a low moisture content (perhaps 10–20 percent or less) will be rigid and hard, while foods with a higher moisture content (perhaps 35 percent or more) will be flexible and soft. Proteins contain nitrogen atoms, and polysaccharides contain oxygen atoms. Both of these atoms are electronegative and therefore form hydrogen bonds with the hydrogen atoms in water molecules, resulting in water molecules being adsorbed to the surface of proteins and polysaccharides. The adsorbed molecules of water lower the Tg of the proteins and polysaccharides, making them more flexible at room temperature and above. In dry wheat flour, the key proteins that form gluten, gliadin and glutenin, are rigid and inflexible at room temperature. When water is added, these two proteins become flexible at room temperature and are able to unfold and move about and to bond with each other to form gluten. Kneading the dough helps to move the proteins around even more, ensuring enough bonding (cross-linking) to form a strong gluten network. The same is true of the starch polysaccharide molecules, amylose and amylopectin. When water is added to starch, the water begins to hydrogen bond with the starch molecules. In this case (usually at 35 percent or more moisture), the Tg of the starch molecules is above room temperature, so it takes heat to reach a temperature at which the granules of starch absorb water and swell. This is known as the *gelatinization temperature* of starch. When dry pasta is cooked, the gluten is hydrated, starch granules absorb water, the Tg decreases, and the pasta changes from hard and brittle to flexible and chewy.

REFERENCES

Collins, J. C. *The Matrix of Life—A View of Natural Molecules from the Perspective of Environmental Water.* East Greenbush, NY: Molecular Presentations, 1991.

Coultate, T. "Water." Chap. 13 in *Food: The Chemistry of Its Components.* 6th ed. Cambridge, UK: Royal Society of Chemistry, 2016.

WHAT IS TEMPERATURE, AND HOW DOES IT DIFFER FROM HEAT?

Temperature is a measure of the *average* kinetic energy (the energy of motion) of all the molecules in a substance. Because it is an average of the kinetic energy of all the molecules, the temperature of a substance does not depend on the total number of molecules in a substance. On the other hand, heat is a measure of the total kinetic energy contained in all of the molecules in a substance, so it is a function of the number of molecules. Heat is a form of energy, while temperature is a measure of the degree of hotness of a substance and is measured by a thermometer. Heat and temperature are not the same.

Kinetic energy (K) is equal to half the mass (m) of a molecule (or a body of some type) times the square of its velocity (v): $K = \frac{1}{2} mv^2$. Therefore the kinetic energy of a molecule is dependent on both the mass of the molecule and its velocity. Because velocity is squared, the kinetic energy of a molecule will be much more dependent on the velocity of the molecule—simply because v^2 will be a much larger number than m. In this sense, temperature can be said to be a measure of the average speed at which all of the molecules in a substance are traveling. This is how temperature is often defined in simple terms (as the speed of the molecules).

Here is an example that distinguishes heat and temperature. A small, red-hot tack is much hotter than a large bowl of warm water. But the warm water contains more heat than the tack. The red-hot tack has a higher temperature than the warm water because temperature is a measure of the degree of hotness of a substance as read on a thermometer. Heat is a form of energy. When a substance is hot, its molecules move more rapidly than when it is cold. The heat of a substance is a measure of the total of all the energy of motion of all of its molecules. Its temperature is a measure of the average energy of motion of its molecules. Because there are many more molecules in the bowl of water than in a small tack (the water weighs much more than a small tack), the water contains more energy in all of its moving molecules. But the tack is at a much higher temperature because the average energy of all of its molecules is greater than that of the molecules in the water.

Distinguishing between heat and temperature is important in cooking science. The cooks at America's Test Kitchen devised a simple test to demonstrate the difference. They carefully heated containers of equal volumes of olive oil and water to 165°F (74°C) in a sous vide water bath and then placed a raw egg in each one. The water turned the egg white cloudy very quickly, indicating the proteins were denaturing and coagulating. The egg in the olive oil looked like nothing had happened—the egg white was still transparent. The experiment is reproduced in figure 1.

Water cooks the egg faster than olive oil because it takes more energy to heat water to 165°F than it does olive oil. The amount of energy required to raise the temperature of a substance is called the *heat capacity* of that substance, and each substance has a different heat capacity. The heat capacity of water is about 2.1 times greater than that of olive oil. All that extra energy required to raise the temperature of water to 165°F is available to cook the food faster. To learn more about heat capacity and cooking food, read the essay in chapter 3 on "Thomas Keller and the Science of Butter-Poached Lobster, and "Why Foods Cook Slower in Oil," *Cook's Illustrated* (March & April 2012): 30.

FIGURE 1

(a) A sauté pan containing olive oil is heated to 165°F (74°C) and a raw egg added moments before the photograph was taken. Notice the egg proteins are transparent and have not started to coagulate. (b) A sauté pan containing water is also heated to 165°F (74°C) and a raw egg added and the photograph taken at exactly the same time as for the pan containing olive oil. Notice the egg proteins are translucent and starting to coagulate, showing the egg is cooking faster in water than in olive oil at the same temperature. Photographs by Daniel J. van Ackere, America's Test Kitchen. Reproduced with permission of America's Test Kitchen.

REFERENCE

Resnick, R., and D. Halliday. "Heat and the First Law of Thermodynamics." Chap. 22 in *Physics for Students of Science and Engineering*, Part 1, 466–488. New York: Wiley, 1962. (This was my first college physics book.)

Once the appropriate vessels were invented, frying food in vegetable oils became very popular. And that may be a good thing because vegetable oils like olive oil are rich in healthy monounsaturated fatty acids and antioxidants. The use of olive oil may also have encouraged the development of seed- and nut-based oils, which are richer in healthy polyunsaturated fatty acids. These would have been a beneficial addition to a diet containing high levels of saturated fats from animals (although palm oil contains fairly high levels of saturated fatty acids, it also contains good amounts of mono- and polyunsaturated fatty acids). Through millions of years of evolution, the human body never developed the enzymes needed to make the two essential polyunsaturated fatty acids required in our diet, linoleic acid and linolenic acid. These omega-6 and omega-3 fatty acids, respectively, are required for the production of a family of extremely important compounds called *eicosanoids*, which play many significant roles in the body. In fact, no mammals produce the enzymes needed to make these two fatty acids. We are dependent on plants and vegetable oils (and fat from animals that eat plants) as our sources of these two essential compounds. This may be a consequence of the diet of very early prehumans: they ate mostly plants, which provided ample amounts of the two essential fatty acids, so there was no need to develop the enzymes for making linoleic and linolenic acids.

FIGURE 2.3

An Akkadian tablet with the first written recipes, dated about 3,750 years ago. Reproduced with permission from the Yale University Babylonian Collection.

Great changes had occurred in agriculture, food, and cooking by the end of the Neolithic period and the start of the Bronze Age, about 4,000 years ago. In the Babylonian collection at Yale University, there is a series of clay tablets known as the Akkadian tablets that includes two tablets with the oldest known cooking recipes, dated 3,750 years ago (figure 2.3). The tablets originated in the area between the Tigris and Euphrates Rivers known as Mesopotamia. The recipes confirm the amazing abundance and

(CONTINUED ON PAGE 36)

RECIPE 2.1 Bone-In Pork Spareribs with Hoisin Barbeque Sauce

INGREDIENTS FOR SAUCE:

2 tsp. peanut oil

2 garlic cloves, finely chopped

¼ cup hoisin sauce

1 Tbsp. light soy sauce

3 Tbsp. sake

1 Tbsp. ketchup

1 Tbsp. rice vinegar

¼ tsp. roasted sesame oil

1½ pounds bone-in pork spareribs

***Yield*: 2 servings**

Some of the foods I most enjoy eating are based on Chinese-style cooking because the ingredients produce complex flavors that are savory, sweet, salty, tangy, and fragrant all at the same time. I have traveled to China, so I have a sense of what the real food is like, but I have to admit that I am not very good at cooking Chinese-style food at home. Pork is the most consumed meat in China, so there are many ways of cooking pork with flavorful savory sauces that provide lots of intense umami taste. The recipe for a Chinese-style hoisin barbeque sauce was developed a number of years ago by barbeque chef Steven Raichlen and adapted here for use in preparing delicious bone-in pork spareribs, but it also goes well with other cuts of pork, chicken, and even beef. It is very simple and almost foolproof, but note that the ribs need to be covered with the barbeque sauce and refrigerated at least 6 hours before cooking. Although the origin of hoisin sauce is not known with any certainty, it is most commonly used with food prepared in the Shanghai region of China. To tell the truth, I like these ribs even more than American-style barbeque pork ribs.

A word of caution is in order regarding many of the soy-based ingredients used in Chinese cooking: they can be very high in sodium and, like this recipe, also very high in sugar. One of the leading brands of hoisin sauce produced in the United States contains approximately 1 gram of sodium and 20 grams of sugar in just 2 tablespoons of sauce. Keep in mind that sodium amounts to about 40 percent of the weight of salt (because one molecule of sodium chloride contains one atom of sodium and one atom of chlorine), so 1 gram of sodium is contained in 2.54 grams of salt (about 0.4 teaspoon); this means 2 tablespoons of the hoisin sauce mentioned will contain almost ½ teaspoon of salt. The U.S. Department of Agriculture guidelines have established an upper limit of 2.3 grams per day of sodium (contained in 5.8 grams of salt, or almost 1 teaspoon), with even lower limits for some at-risk groups with elevated blood pressure. So read the Nutrition Facts label on soy-based products for guidance, and follow the simple rule of eating all foods in moderation, in both quantity and frequency, rather than abstaining altogether.

These ribs go well with dry-cooked spicy string beans, jasmati rice, and bok choy sautéed with chopped garlic in peanut oil. To prepare enough beans for two, cook about two cups of them on medium heat in a small amount of hot peanut oil in a covered skillet until well browned but still relatively firm (about 5 minutes). Remove the pan briefly from the heat, and add a mixture of 1 tablespoon of light soy sauce, 1 tablespoon of Shaoxing cooking wine, and 1 teaspoon of commercial chili garlic paste (less if you don't want them too spicy). Mix well to coat the beans. Cover the skillet, and finish cooking the beans on medium heat for about 5 more minutes, until the beans are relatively tender.

DIRECTIONS:

First, prepare the hoisin barbeque sauce. In a small saucepan, cook the oil and garlic on medium heat until the garlic is very lightly browned, about 1–2 minutes (depending on the heat). Do not allow the garlic to brown too much, or it will become bitter. Add the hoisin sauce, soy sauce, sake, ketchup, and rice vinegar, and continue cooking the mixture until the sauce is reduced by about 25 percent and starts to thicken; then allow the sauce to cool, and mix in the sesame oil.

Next, prepare the ribs. Pour about a quarter of the cooled sauce in the bottom of an ovenproof casserole dish large enough to hold the ribs. Then place the ribs on top of the sauce, and pour the remaining sauce evenly over the top of the ribs. Cover with plastic wrap, and refrigerate the ribs for at least 6 hours.

When ready to cook the ribs, remove them from the refrigerator, and allow them to warm to near room temperature (about 20 minutes); replace the plastic wrap with aluminum foil. Meanwhile, heat the oven to 325°F. Place the covered casserole dish on the middle rack, and bake the ribs for 1½ hours. Uncover the ribs, and continue baking for another 30 minutes to concentrate the sauce. (Don't cook the ribs any longer than 2 hours total, or they will become too dry.) Remove the ribs from the excess sauce, and serve.

variety of foods available at that time. Dishes described in the recipes include a great variety of meats, fowl, fish and shellfish, grains, vegetables (including root vegetables), mushrooms, many varieties of fruit, honey, milk, butter, animal fats, vegetable oils (including sesame and olive oils), numerous kinds of bread, beer, and wine. On one tablet, there are twenty-five recipes for stews, twenty-one with meat and four with vegetables, showing an amazing richness and refinement. The stews contain a wide variety of meats and some vegetables, are slowly cooked in covered clay pots, and are seasoned with a variety of herbs and spices, as well as beer and wine. The ancient Egyptians and Greeks, and later the Romans, took advantage of the great abundance of readily available food by holding huge banquets for the ruling classes.

Roughly 2,500 years ago, the Chinese discovered the process for fermenting salted soybean paste as a method for preserving soybeans. The fermented paste found its way as a seasoning to Japan where, over many years, it evolved into Japanese soy sauce, known as *shoyu*. About the same time (2,700–2,500 years ago), the Chinese developed a process for making cast iron from raw iron ore, and relatively soon afterward, during the Golden Age of the Han Dynasty (2,200–1,800 years ago, when classical Chinese cuisine evolved), they created the first cast-iron wok for frying food with very hot sesame oil. Almost immediately, the Chinese developed the technique of stir-frying food in a very hot wok, which the famous Chinese chef Eileen Yin-Fei Lo describes as a process of cooking food to the precise point of doneness—and no more. Known as *ts'ui*, "it is a quick process and, when done properly, foods are never overdone, their fabric is never destroyed, their taste remains sharp and in focus." To accomplish this, all of the food must be cut into bite-sized pieces and ready to tip into the wok at the precise moment, depending on the food. This surely set Asian cooking on a path all its own.

The First Scientific Theories Evolve

We have finally reached the point where ancient Greek scholars are beginning to develop theories about the nature of matter using the scientific method of forming hypotheses, designing tests, and answering questions based on observable facts. Without these theories, it is not possible to think about cooking in scientific terms. The earliest theories originated with the Greek philosopher Thales of Miletus about 2,600 years ago. He was concerned with explaining how the entire universe worked, including the states of matter, for which he invoked the concept of elements, although not the kind of elements we think of from the periodic table. Thales and his fellow philosophers conceived of the elements as earth, water, air, and fire, representing solid, liquid, gas, and heat, respectively. The four elements were interchangeable, as water can be converted to a solid (earth) by freezing to make ice or to a gas (air) by heating with fire to produce steam. It fell upon Aristotle to finally unite the various theories and synthesize a cohesive concept around 2,300 years ago. Aristotle explained that everything we can see and touch is produced by the combination of matter and form. Each form consisted of the opposing qualities of heat, cold, moisture, and dryness, thus giving qualities to each of the four elements. Earth had the qualities of cold and dryness, water the qualities of cold and moisture, air the qualities of heat and moisture, and fire the qualities of heat and dryness. The elements were now interchangeable. Building on the ideas that the pre-Socratic Greek philosopher Democritus had formulated about 100 years earlier, Aristotle envisioned the four elements as constructed of what he called "compound atoms," the actual building blocks of all matter. Amazing! Aristotle and Democritus actually conceived of atoms long before they were proven to exist. It was not until 1805 that John Dalton provided a precise proof of the concept of atomic theory. When William Blake wrote the beautiful opening lines of his poem "Auguries of Innocence" in 1803, he must have known of Aristotle's concept of compound atoms as the building blocks of all matter, including a "Grain of Sand" and a "Wild Flower." He even imagined these atoms were so small that an infinite number could fit "in the palm of your hand" (the opening four lines of the poem are quoted in the preface to this book).

Aristotle's contributions to scientific theory were so powerful they dominated scientific thought for many centuries. In one sense, this was a great step forward, but in another, it limited the conception of new ideas and led many scholars in the wrong direction. If matter was interchangeable, then gold, the most valuable of all metals, could be formed from other, less desirable metals. The "science" of alchemy was born. Alchemists were convinced that mercury and sulfur could be combined to produce gold, and so the quest began. During the medieval period (from the fifth to the fifteenth century), science fell under the influence of alchemists and took a hiatus as far as cooking was concerned. To be sure, there were advances in science. For example, Ibn al-Haytham (965–1040), the great Arab mathematician and physicist, was credited with clearly defining the modern scientific method of experimentation and also recognized that the human eye sees a cone of rays from an object, thus giving it a three-dimensional perspective. But cooking in that period underwent only modest changes, including the incorporation of spices from Asia, the use of sugar to sweeten and preserve foods, and the development of new methods for preserving food by salting, pickling, and drying—but nothing equivalent to what began to evolve in the 1500s.

The quest to transmute substances into gold lasted until about the sixteenth century, when irrefutable facts finally caught up with the last of the alchemists. New thinkers were coming up with new scientific theories that changed the understanding of how the world works. In 1661, Robert Boyle finally demolished Aristotle's concept of the four elements when he proceeded to experiment with combustion as the basis of fire, proving it was a process rather than an element. Between 1774 and 1777, Joseph Priestley and Carl Wilhelm Scheele independently discovered oxygen as the component of air responsible for combustion. And Antoine-Laurent Lavoisier pulled everything together by proving the role of oxygen in combustion in 1789. His carefully designed and beautifully executed experiments turned chemistry into a true science, and as a result, he is known as the father of modern chemistry. Unfortunately, his brilliant contributions to science ended in 1794 when he was executed by the guillotine during the French Revolution.

Finally, no review of cooking science could be complete without an understanding of heat. Until 1798, scholars of that era still believed heat was a physical substance like water or mercury. That same year a brilliant American from Massachusetts named Benjamin Thompson, and later known as Count Rumford of Bavaria, provided the first real evidence that heat was a form of energy and not a physical substance. While boring cannons for the Bavarian government, he concluded the heat generated during boring was caused by mechanical friction and could not be a physical substance as everyone had believed. Science was finally revivified, heading in a direction that would bring cooking to the highest levels of gastronomic accomplishments over the next several hundred years. As we will learn in chapter 4, it was no coincidence that French cooking reached its zenith with Antoine Carême's nouvelle cuisine shortly after French science achieved similar status with Lavoisier's exquisite experiments.

3 Early Science Inspires Creativity in Cooking (1500–1799)

Science Is Reborn During the Sixteenth and Seventeenth Centuries

The early pioneers of well-planned, carefully executed scientific experiments, such as Robert Boyle and Antoine-Laurent Lavoisier, provided the foundation for the classical sciences of physics, chemistry, and biology as early as the sixteenth century. But unlike the pure sciences, cooking science does not have a history of academic research on which to build a knowledge base. Even food science is a relatively new field. The first formal academic department of food science that carried out scientific research was established at the University of Massachusetts at Amherst in 1918, with Dr. Walter Chenoweth as its head. Claimed to be the first in the United States, it may also be among the first departments of food science in the world.

Cooking science is an applied science based on the knowledge of physics, chemistry, biology, engineering, nutrition, and, of course, food science. Until very recently, there were no academic departments dedicated to cooking science; rather, the closest academic field was gastronomy, which involves the study of food history, culture, and both classical and modern cooking techniques known as molecular gastronomy. Two good examples are the gastronomy programs at Wageningen University in the Netherlands and at Boston University in the United States. In addition, scientific journals for publishing research on cooking science have only recently been introduced. The *Journal of Culinary Science and Technology* has been published since 2005, and the *International Journal of Gastronomy and*

FIGURE 3.1

The Alchemist, by Pieter Bruegel the Elder, 1558, from the collection of the Metropolitan Museum of Art, New York. Note how the activities and expressions on the faces of the alchemists mock the "science" of alchemy.

Food Science, created in 2012, publishes peer-reviewed research in cooking science. The Department of Public Health at the University of Parma in Italy has published extensive research on the impact of cooking on nutrients in food (see chapter 6). Recognizing the need to apply science to the culinary arts, the Culinary Institute of America initiated a bachelor's degree program in culinary science in 2015. The founders of this program realized that chefs working in commercial kitchens benefited from knowledge of cooking science, allowing them to create and scale up recipes without sacrificing the consistently high quality of dishes made with ingredients that may vary in composition, such as their water, sugar, or starch content, depending on how they were harvested and stored. As an example, a local food processor I recently visited told me that, when making potato salad, they have to adjust the cooking parameters each time they change from using potatoes grown in Massachusetts to using potatoes grown in Maine. They have learned that producing the same quality batch after batch benefits from the application of cooking science

Let's continue the evolutionary journey of cooking science by going back more than six centuries from the present to learn how the early theories of the classical sciences first evolved and how they were applied to cooking. As previously mentioned, alchemy dominated scientific thinking as far back as 2,000 years ago when it was practiced in the ancient city of Alexandria in Egypt. From there, it spread to the Far East and medieval Europe, surviving until the early 1700s. One of the first people to challenge alchemy was the Italian metallurgist Vannoccio Biringuccio (1480–1539). His writings did not involve food or cooking, but in the following quote from his book *De la Pirotechnia*, published in Italian (not Latin) shortly after his death in 1540, he strongly condemned alchemy, paving the way for new thinking about chemistry and science:

> For this reason I tell and advise you that I believe that the best thing to do is turn to the natural gold and silver that is extracted from ores rather than that of alchemy, which I believe not only does not exist, but also in truth, has never been seen by anyone, although many claim to have seen it. For it is a thing whose principles are unknown, as I have told you; and whoever does not know the first principles of things is even less able to understand the end.

Because we know the ultimate fate of alchemy, Biringuccio's statement seems very logical and insightful today. But how was his criticism viewed at the time when alchemy was an accepted science? Was he ridiculed for being a heretic or praised for being a visionary?

THE DIFFERENCE BETWEEN WAXY AND MEALY POTATOES

No doubt you have read that high-starch mealy potatoes, like Russet Burbank, are best for mashing and baking, while low-starch waxy potatoes, such as Red Bliss, are best for boiling and making potato salad. Much has been written about why waxy and mealy potatoes perform differently, but not all of it is consistent with the research on the cooking properties of different varieties of potatoes. So let's see what the research tells us.

Let's start by defining what we mean by waxy and mealy potatoes. According to Professor Diane McComber (Iowa State University, retired), sensory panelists describe cooked waxy potatoes as moist, mushy, and smooth, while cooked mealy potatoes are harder and drier and produce a sensation of particulate matter in the mouth. Waxy potatoes tend to have a thin skin, be less dense, and contain less starch (about 16 percent on a wet weight basis) and more moisture. Mealy potatoes tend to have a thicker skin, be more dense, and contain more starch (about 22 percent on a wet weight basis) and less moisture. On a molecular level, the starch in waxy potatoes is composed almost entirely of the large branched molecule called *amylopectin*, while the starch in mealy potatoes is composed of a mixture of amylopectin molecules (about 74 percent) and the much smaller, linear *amylose* molecule (about 26 percent). Both amylopectin and amylose are polysaccharides, which function as storage forms of glucose.

Further research by McComber showed that, when high-starch Russet Burbank potatoes were cooked (steamed), the potato cells became "completely engorged with gelatinized starch," in contrast to two varieties of low-starch waxy potatoes, which appeared to be "only 30–50 percent filled" with swollen starch granules. Using nuclear magnetic resonance spectroscopy, the starch granules, present at a higher level in Russet Burbank potatoes, were observed to absorb more moisture, while less of the moisture in the lower-starch waxy potatoes was absorbed by the swollen starch granules, leaving more free moisture. This explains why mealy potatoes are perceived as dry, while waxy potatoes are characterized as moist. On eating, the waxy potatoes release the loosely held water that is not bound up by the lower level of gelatinized starch.

Interestingly, McComber's research showed that in both the waxy and the mealy varieties, the cells of steamed potatoes were intact and not collapsed when visualized by scanning electron microscopy. Her research also showed that calcium and magnesium ion concentrations were higher in Russet Burbank potatoes than in the Pontiac variety of waxy potatoes. Calcium and magnesium ions are known to strengthen pectin, the polysaccharide that is part of the cell wall structure and, more importantly, that acts as a glue to hold the cells together. These observations led her to agree with earlier research that had concluded the cells of steamed mealy potatoes resist separation into individual cells but break "into particulate masses," producing a texture that is less smooth than that of waxy potatoes.

This research goes a long way in explaining the differences in texture of mealy and waxy potatoes, but it doesn't entirely explain why they behave differently when cooked. To help shed some light on this question, Joseph Bazinet, a student in my 2011 Food Analysis class at Framingham State University, undertook a project to measure the cooking differences of three varieties of potatoes using a texture analyzer instrument manufactured by Brookfield Engineering (see figure 6.2). His research helps to explain some of the cooking differences of waxy and mealy potatoes. For the study, we chose high-starch mealy

Russet Burbank, low-starch waxy Red Bliss, and intermediate Yukon Gold, which is described as a medium-starch potato with cooking properties in between waxy and mealy potatoes. Small cylinders (1.3 × 2.5 centimeters) of each variety were removed and cooked in boiling distilled water for 10 minutes; they were then plunged into ice water to stop further cooking. The force (measured as the peak load in grams) needed to compress each cylinder by 40 percent was recorded with the texture analyzer, using a flat disk with a diameter wider than the potato cylinder (numerous conditions were tested before selecting this method as being the most reliable). The results are shown in figure 1 (based on nine replicates of each variety).

From these results, it is clear that mealy Russet Burbank potatoes required the least amount of force to compress the boiled potato cylinders. Yukon Gold and Red Bliss required similar amounts of force, with perhaps a slightly lower force required for the intermediate Yukon Gold potatoes. These results indicate the cell structure of mealy potatoes is more readily broken down by boiling water and are consistent with the observation that mealy potatoes are better suited for mashing and baking. Both waxy Red Bliss and intermediate Yukon Gold hold their shape when boiled and thus are better suited for use in potato salad. These results also suggest that Yukon Gold potatoes behave more like waxy potatoes. However, these results are from only one lot of commercial potatoes, and it is well known there is significant variation in density and starch content even between potatoes of the same variety.

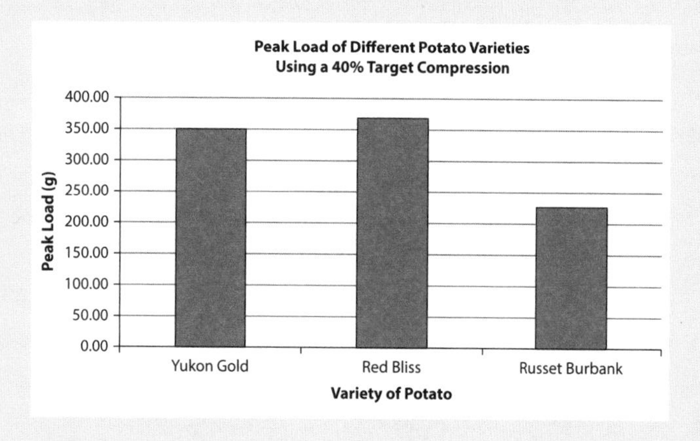

FIGURE 1

A graph showing the force required to penetrate three varieties of potatoes cooked under identical conditions, measured using a texture analyzer (see figure 6.2). Note that it takes less force to penetrate the Russet Burbank potato than the Yukon Gold and Red Bliss potatoes. The graph is drawn from the author's research at Framingham State University in 2011.

Research from the Agrotechnological Research Institute, Wageningen, Netherlands, supports our results. Researchers there found that more pectin was released when mealy potatoes were boiled compared with waxy potatoes. Furthermore, they concluded from transmission electron microscopy that the release of more pectin from the cooked mealy variety of potatoes resulted in cell wall loosening and increased cell sloughing (most likely clumps of cells, as suggested by McComber) compared with waxy potatoes. Again, this is quite consistent with the experiment showing less force is required to compress cooked mealy potatoes, making them better for mashing and baking but less smooth and moist.

More recent research published by the same group from the Netherlands convincingly showed that the dry matter (DM) content of mealy, waxy, and intermediate potato varieties strongly overrules any cultivar-specific effects on the cooking properties of potatoes and that the DM directly correlated with the starch content of each variety. *Thus, the starch content of a potato variety has the greatest impact on cooking properties.* Not only do mealy potatoes contain more starch, but also the composition of the starch is different (a mixture of amylose and amylopectin), as mentioned earlier. As heat penetrates a cooking mealy potato, the starch granules swell and gelatinize more rapidly than in waxy potatoes. The greater pressure exerted by large numbers of swelling starch granules combines with the heat that is breaking down the pectin glue between the potato cells to force the cells to swell, rupture, separate, and slough off more readily during cooking.

All this research just to understand why Russet potatoes make better mashed potatoes! And we knew this all along!

REFERENCES

McComber, D. R., H. T. Horner, M. A. Chamberlin, and D. F. Cox. "Potato Cultivar Differences Associated with Mealiness." *Journal of Agricultural and Food Chemistry* 42 (1994): 2433–2439.

McComber, D. R., E. Osman, and R. Lohnes. "Factors Related to Potato Mealiness." *Journal of Food Science* 53 (1988): 1423–1426.

Nonaka, M. "The Textural Quality of Cooked Potatoes. I. The Relationship of Cooking Time to the Separation and Rupture of Potato Cells." *American Potato Journal* 57 (1980): 141–149.

Trinette van Marle, J., K. Recourt, C. van Dijk, H. A. Schols, and A. G. J. Voragen. "Structural Features of Cell Walls from Potato (*Solanum tuberosum* L.) Cultivars Irene and Nicola." *Journal of Agricultural and Food Chemistry* 45 (1997): 1686–1693.

Van Dijk, C., M. Fischer, J. Holm, J.-G. Beekhuizen, T. Stolle-Smits, and C. Boeriu. "Texture of Cooked Potatoes (*Solanum tuberosum*). 1. Relationships Between Dry Matter Content, Sensory-Perceived Texture, and Near-Infrared Spectroscopy." *Journal of Agricultural and Food Chemistry* 50 (2002): 5082–5088.

Another important figure in the evolution of cooking science is Jan Baptist van Helmont (1577–1644), a Flemish chemist who also earned a doctor's degree in medicine. He was the first to coin the name *gas* (perhaps around 1635–1640) when he observed what he described as "gas sylvestre," the gas produced by fermenting grape must, which he also called the "gas of wines." Van Helmont noted that it was the same gas produced by burning charcoal and that is was distinctly different from air, although he had no idea that it was actually carbon dioxide. He also studied digestion and was the first to suggest that food is digested in the body by what he called "ferments," which today would be considered the same as digestive enzymes. Although Van Helmont believed in alchemy and asserted that the only true elements were air and water, he at least contributed a few original scientific ideas that still hold true today.

Next we come to the contributions of Robert Boyle (1627–1691), who published more than twenty scientific papers and books, including his most famous, *The Sceptical Chymist*, in 1661. He did most of his experiments on the nature of the elements and the role of air in combustion with Robert Hooke (1635–1703), his faithful assistant for many years at Oxford University. In *The Sceptical Chymist*, Boyle challenged Aristotle's four elements—earth, air, water, and fire—by defining elements as "unmingled bodies not being made of any other bodies," which is consistent with our present-day definition of the pure elements within the periodic table, like copper and sulfur. In 1650, Boyle learned of the work of Otto von Guericke (1602–1686) in Germany, which included his experimenting with air pressure and vacuum and designing a pump to remove air from within closed vessels. Boyle repeated von Guericke's work in 1660 and designed his own pump with which he could remove air from sealed glass containers, enabling him to study the effect of air on combustion. In his 1672 work "New Experiments Touching the Relation Betwixt Flame and Air," Boyle described numerous experiments in which he burned various substances such as sulfur in both the presence and the absence of air. He was able to prove with careful experimentation that air was an essential component of combustion and that fire was not one of the four basic elements. The scientific principles established by Aristotle several thousand years earlier were finally beginning to crumble, making way for a new renaissance in science.

The Rebirth of Science—Now We're Really Cooking!

Boyle also experimented with air and pressure, leading to his formulation of Boyle's Law: the pressure of a gas, such as air, increases as the volume of the gas decreases

FIGURE 3.2

Dennis Papin's pressure cooker, 1680–1681. Notice the long weighted arm on the left for releasing pressure. Photograph by P. Faligot. From the photographic collection of Musée des Arts et Métiers, Paris. Reproduced with permission.

(that is, compressing a gas into a smaller volume increases the pressure of the gas). He published his research in 1662, and several years later it caught the attention of Denis Papin (1647–1713), a young French physicist and mathematician. Papin visited London to meet with Boyle in 1675, and the two got along so well that Papin stayed and worked with Boyle from 1676 to 1679. Based on Boyle's Law, Papin concluded that heating water in a closed container with a fixed volume would cause the pressure of the water to increase, allowing the boiling point of water to be increased above its normal boiling point of 212°F (100°C) at atmospheric pressure. Unfortunately, even a heavy metal container would explode from the pressure of the steam if it became too hot! As a result, Papin designed the first safety valve, using a weight to open a valve and release the pressure before it got high enough to explode the container. This clever invention led Papin to design his "steam digester" for cooking food under pressure at temperatures higher than the normal boiling point of water, thus creating the first pressure cooker, which he presented to the Royal Society of London in 1679 (figure 3.2). It is reported that Papin's steam digester was used to cook inexpensive fatty cuts of meat and bones in water to produce delicious stews. After leaving London, Papin moved to Germany, and in 1690, he built the first model of a piston-driven steam engine, which he used to power a paddle-driven boat. Unfortunately, life was not very kind to Papin; he died a pauper and was buried in a pauper's grave in 1713.

The new experimental science developed in Boyle's era began to have a significant influence on seventeenth-century French cooking. A very innovative chef named François Pierre de La Varenne (1615–1678) published the first influential cookbook on French cooking, *Le Cuisinier françois*, in 1651. His book contains the first mention of preparing a stock, known as a *fonds de cuisine*, in which he recommended using discarded mushrooms as the base for the stock rather than throwing them out. La Varenne cleverly used egg whites to clarify his stocks by allowing the albumin proteins in egg whites to bind with the insoluble proteins from meat that cause the stock to become cloudy.

MAKING STOCK OF THINGS

French chef François Pierre de La Varenne (1615–1678) is credited with publishing the first recipe for a simple stock to flavor a dish in his 1651 cookbook, *Le Cuisinier françois*. There the frugal La Varenne describes how to make mushroom stock from blemished mushrooms rather than discarding them. It was more than 150 years later that the great French chef Marie-Antoine Carême (1784–1833) perfected the basic process still used today of soaking meat in cold water and then slowly warming and simmering the mixture to avoid overly coagulating the protein while concentrating the relatively clear stock. The French have been masters at making great stock ever since.

Carême's procedure can be broken down into three basic steps: extraction, filtration, and reduction. The first step extracts flavor molecules from the ingredients with water or wine, while the second step removes unwanted coagulated protein, fat, and insoluble ingredients to produce a clear stock. The last step evaporates some of the water to produce a concentrated, more flavorful stock. Each of these steps plays a role in creating the flavor of the stock. The ingredients determine the specific flavor molecules that are extracted, while heating time and temperature control both the loss of volatile aroma compounds and the formation of new flavor molecules. Filtration not only produces a clear stock but also removes undesirable bitter tastes present in coagulated protein (scum) and rancid fat. The reduction step concentrates flavor and provides additional time for the creation of more flavor molecules. In addition to creating flavor, the heating steps (extraction and reduction) release gelatin from the collagen in the connective tissue of meat and bones. Gelatin provides viscosity and a pleasing unctuous mouthfeel to the stock.

A traditional stock usually calls for meat (or fish), bones, and vegetables, typically onions, carrots, and celery. Raw meat, bones, and vegetables may be placed in cold water to start the extraction process, or they may be roasted in the oven prior to extraction to enhance the flavor of the stock. The process starts with extracting the molecules we taste and smell from the meat, bones, and vegetables. During the optional oven-roasting step, nonvolatile nucleotides are derived from the energy-storing compound adenosine triphosphate (ATP), proteins break down to nonvolatile peptides (small fragments of proteins) and even smaller amino acids, and onions and carrots release nonvolatile sugars. Because they are nonvolatile, we cannot smell any of these molecules. But they are all soluble in water, so they can be extracted and tasted in the stock. The peptides and amino acids combine with the nucleotides to produce a potent savory, meaty umami taste, while the sugars and an amino acid called *glycine* lend sweetness. Under the conditions of dry heat in the oven, volatile aroma molecules that we can smell are also produced. The peptides and amino acids react with certain of the sugars (called reducing sugars, such as glucose and fructose but not sucrose) by the Maillard reaction to produce very potent, volatile aroma molecules (for more on the Maillard reaction, see chapter 5). These include the caramel-like 4-hydroxy-2,5-dimethyl-3(2H)-furanone, the popcorn-smelling 2-acetyl-1-pyrroline, and the meaty-smelling sulfur-containing 3-(methylthio)propanal. In addition, the fats are oxidized to compounds with a deep-fried odor, such as 2,4-decadienal and 2,6-nonadienal. These two compounds are soluble in fat but not water. During the slow extraction step in simmering water, the onions (or leeks) produce a sulfur-containing water-soluble compound called 3-mercapto-2-methylpentane-1-ol (MMP). In 2011, German researchers showed that the strong meaty aroma of this compound is the most potent contributor to the flavor of stock, along with 2,4-decadienal, mentioned above. But MMP is formed only when the onion is finely chopped, not when it is left whole. Don't forget to chop the onion!

After slowly simmering the stock for many hours, any protein scum that has floated to the top is removed by skimming, and the extract is filtered until clear. The last step in the process is reduction. Most cookbooks call for reducing the stock to a specific *volume* to concentrate the water-soluble, nonvolatile taste molecules rather than specifying *how long* the stock should be heated. During this step, some of the volatile aroma compounds are lost with the escaping steam. At the same time, additional aroma compounds are slowly formed. This is especially true of the very important compound MMP, which takes several hours to form in significant amounts. Meanwhile the proteins continue to break down to peptides and amino acids, especially umami-tasting glutamate, while nucleotides continue to form as a result of the ATP breakdown. This is very important because recent research from Denmark has shown that flavor is affected not only by the reduction volume but also by the length of time the stock is heated. And the two are not directly related. Depending on the amount of heat applied (high or low flame), the stock can be reduced rapidly or slowly. Fast reduction drives off some aroma compounds without allowing sufficient time to generate new flavor molecules. Compounds like MMP and 2,4-decadienal require a certain amount of time to form, so a slow simmer over a longer period of time produces more of these crucial compounds. Finally, a study conducted at the Culinary Institute of America determined the optimum temperature and time to form peptides and amino acids from proteins and nucleotides from ATP: the stock should be reduced at 185°F (85°C) for a minimum of 60 minutes. The flavor of the stock clearly changes with time, so it is advisable to reduce the stock slowly for several hours.

Like flavor compounds, gelatin is formed very slowly, by the breakdown of collagen in the connective tissue of meat and bones. Collagen breakdown takes hours at relatively low temperatures. At around 185°F, the temperature at which water slowly simmers, it takes about 6 hours for significant amounts of gelatin to form. Therefore, the extraction and reduction steps are best conducted over a period of at least 6–8 hours to increase the formation of gelatin and important flavor molecules such as MMP. It is not uncommon to heat a stock for 16–20 hours, although it is unclear if this much time is necessary. But it does explain why slowly reducing a stock to 25–50 percent of its original volume produces such a flavorful concentrated demi-glace. It is far more than just the reduction in volume.

Several other factors are important. When the extraction step is started with cold water, the proteins slowly coagulate and float to the top so they can be easily removed by skimming the surface as well as during filtration. Starting with hot water and boiling the liquid breaks the coagulated protein down into small bits, which are hard to remove and result in a cloudy stock. Finally, veal bones are frequently added in the preparation of both veal and beef stocks. The degree of collagen cross-linking in the connective tissue of bones increases with the age of the animal. Because veal is produced from a young calf rather than an older steer, the collagen in veal bones is much less cross-linked than that in beef bones. This means the collagen in veal bones breaks down more rapidly and completely to gelatin, so veal bones yield a lot more gelatin.

REFERENCES

Christlbauer, M., and P. Schieberle. "Evaluation of the Key Aroma Compounds in Beef and Pork Vegetable Gravies a la Chef by Stable Isotope Dilution Assays and Aroma Recombination Experiments." *Journal of Agricultural and Food Chemistry* 59 (2011): 13122–13130.

Krasnow, M. N., T. Bunch, C. F. Shoemaker, and C. R. Loss. "Effects of Cooking Temperatures on the Physiochemical Properties and Consumer Acceptance of Chicken Stock." *Journal of Food Science* 77, no. 1 (2012): S19–S23.

Snitkjaer, P., M. B. Frost, L. H. Skibsted, and J. Risbo. "Flavour Development During Beef Stock Reduction." *Food Chemistry* 122 (2010): 645–655.

Using his method for making stock, La Varenne was the first to describe the preparation of milk-based béchamel sauce (which doesn't require clarifying), as well as a roux made of pork fat for thickening sauces in place of breadcrumbs. Without realizing it, he had prepared some of the first emulsified sauces. In the case of béchamel sauce, the starch from the roux interacted with the milk proteins to form a stable thickened emulsion. La Varenne may have been the first to prepare a Hollandaise-type emulsion, stabilized with egg yolk as the emulsifying agent, as a sauce for asparagus. Through his influence, stocks became the base for preparing many great sauces, setting the stage for French cooking to become one of the great cuisines of the world in the eighteenth and nineteenth centuries, following the leadership of the celebrated chef Marie-Antoine Carême (1784–1833). Some renowned French chefs of the time, such as François Marin (chef to Madame de Pompadour and author of the three-volume cookbook *Les Dons de Comus*, published in 1742), even described themselves as chemists and cooking as a form of chemistry. But more on these science-guided chefs later when we reach the pinnacle of French science with Antoine-Laurent Lavoisier, just before the French Revolution.

More than 100 years elapsed from the time of Boyle's experiments to the discovery that air contained the pure element oxygen, the actual gas required for combustion, and that the heat produced by combustion was not a substance but a form of energy. Now we're really cooking! Without heat, there can be no cooking in the true meaning of the word (to prepare food for eating by applying heat). The discovery of oxygen is a fascinating story, with two scientists independently discovering the substance in air that makes life possible.

In 1772, Carl Wilhelm Scheele (1742–1786), a Swedish apothecary chemist, became the first to prepare oxygen, although at the time he had no idea what he had discovered. He did this in his pharmacy near Stockholm by heating a variety of inorganic compounds such as mercuric oxide, potassium nitrate, and silver carbonate, and he called the liberated gas "fire air" because a candle would burn brightly in it. The gas was not named oxygen until many years later when its true chemical structure was determined. Unfortunately, Scheele did not feel any urgency about publishing his important discovery. Even though his manuscript was delivered to the publisher in 1775, it was not actually published until 1777. The reason for the delay is not known.

It is common knowledge that oil and water don't mix. If you try to mix them together, they quickly separate, with the water sinking to the bottom and the oil floating on top. If you mix them very vigorously, one of them will break up into droplets and disperse in the other. But even this dispersion won't last long, and the two will soon separate as before.

Vigorously mixing oil and water has two possible outcomes. In one, droplets of oil are dispersed in a continuous phase of water. In the other, droplets of water are dispersed in a continuous phase of oil. The first form (oil droplets dispersed in water) is called an oil-in-water emulsion (O/W emulsion), while the second form (water droplets dispersed in oil) is called a water-in-oil emulsion (W/O emulsion). (See figure 1.)

Distinguishing between an O/W emulsion and a W/O emulsion is very important because the mouth senses only the continuous phase rather than the dispersed phase. Mayonnaise is a perfect example. Mayonnaise contains about 80 percent oil and 20 percent vinegar, plus small amounts of egg and seasoning like mustard and salt. The surprising thing about mayonnaise is that, even with four times more oil than vinegar, the oil is dispersed as tiny droplets in a continuous phase of vinegar (O/W emulsion). As a result, mayonnaise does not feel greasy in the mouth because the mouth senses only the continuous watery vinegar phase and not the dispersed oil droplets.

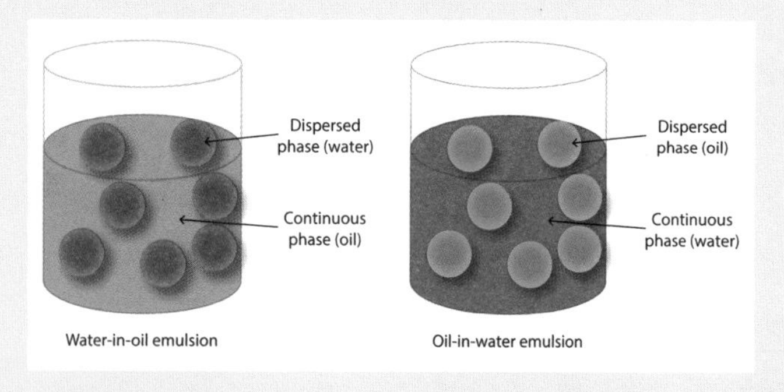

FIGURE 1

The composition of water-in-oil (W/O) and oil-in-water (O/W) emulsions.

Now consider a simple vinaigrette made by vigorously mixing about four parts of oil with one part of vinegar. In this case, the vinegar becomes dispersed as droplets in a continuous phase of oil (W/O emulsion). If the vinaigrette is prepared by slowly adding the oil to the vinegar with very vigorous mixing, the W/O emulsion will usually stay together long enough to taste—and even drizzle on some salad greens. A W/O vinaigrette "tastes" very oily compared with mayonnaise. Another example is butter, which is also a W/O emulsion. Butter feels greasy in the mouth rather than wet.

But why does mayonnaise exist as an O/W emulsion while a vinaigrette, containing the same ratio of oil to vinegar, exists as a W/O emulsion? Confused? The answer is actually quite simple. The mayonnaise contains egg and mustard, which act as *emulsifiers* to stabilize the oil as droplets. Emulsifiers are substances that facilitate the dispersion of one phase (as tiny droplets) into another. A simple vinaigrette does not contain emulsifiers, so the smaller volume of vinegar ends up being dispersed as droplets in a much larger continuous phase of oil. Without an emulsifier, the liquid used in excess usually forms the continuous phase.

Egg yolks contain lipoproteins and phospholipids, like lecithin, that coat the surface of the oil droplets and prevent the droplets from coalescing and forming a continuous phase even though the volume of oil is four times greater than the volume of vinegar. But not all emulsifiers stabilize the oil as droplets. Some are better at stabilizing the vinegar as droplets. It depends on the properties of the emulsifier.

As a general rule, the continuous phase is the one in which the emulsifier is soluble. If an emulsifier is more soluble in oil, then oil will form the continuous phase regardless of the proportions of oil and vinegar. Similarly, water-soluble emulsifiers stabilize vinegar as the continuous phase. In the home kitchen, we have relatively few emulsifiers to pick from—namely, egg yolks, mustard, and casein in milk. Both egg yolks and mustard tend to stabilize emulsions with oil droplets suspended in vinegar. That's why adding a little mayonnaise, which contains egg yolks, to a mixture of three parts of oil to one part of vinegar forms a fairly stable W/O emulsion.

The food-processing industry has literally dozens of emulsifiers from which to choose. They range from water soluble to oil soluble. The solubility properties are expressed in terms of the emulsifier's *hydrophilic-lipophilic balance* (HLB). Hydrophilic substances are water loving, while lipophilic substances are fat (oil) loving. The HLB scale runs from 0 to 20. Emulsifiers with a high HLB value (for example, sodium stearoyl lactylate) are hydrophilic and water soluble. Emulsifiers with a low HLB value (for example, glycerol monostearate) are oil soluble. Emulsifiers with HLB values between 3 and 6 stabilize W/O emulsions, while emulsifiers with HLB values between 11 and 15 stabilize O/W emulsions. Emulsifiers with intermediate values (8–10) are good wetting agents (they promote the spreading of a liquid phase, such as water, onto a solid phase, such as cocoa powder) but relatively poor emulsifiers.

So far I have discussed emulsions stabilized with emulsifiers. That is, the emulsifier prevents the oil and water from separating by forming a protective barrier around the droplets. But there is another way to stabilize emulsions. Thickening agents like starch, flour, and gums also stabilize emulsions, but they are not emulsifiers. They do not form a protective barrier around the dispersed droplets. Instead, these substances increase the viscosity of water. When oil is dispersed in water that has been thickened with starch (the starch-water mixture must be heated first to thicken it), the oil droplets will be stabilized because the high viscosity of the starch-water continuous phase prevents the oil droplets from moving around and coalescing. Can you think of any examples? Soups, sauces, and gravies thickened with cornstarch or flour form stable O/W emulsions with any fat that may be present. The O/W sauce will be creamy and smooth but not greasy because water is the continuous phase. Check the label of those superstable creamy salad dressings in the supermarket, and see if any of them contain starch.

When a stable emulsion is made, it can be difficult to see whether the continuous phase is oil or water. Taste can sometimes give an indication, as described above. But the best way to identify the continuous phase is to measure the electrical conductance. An emulsion with a continuous vinegar phase (plus a little salt) will readily conduct a low-voltage electric current, but an emulsion with a continuous oil phase will not. As an interesting exercise, use this method to check emulsions made with different ratios of oil and water stabilized with different emulsifiers or none at all.

REFERENCE

Stauffer, C. *Fats and Oils*. St. Paul, MN: Eagan Press, 1996.

RECIPE 3.1 Christine's Rich Brown Gravy

INGREDIENTS:

1 medium onion, finely chopped

1 carrot, peeled and cut into small disks

1 celery stalk and some leaves, chopped

1 turkey neck and bag of giblets (gizzard, heart, and liver) provided inside the turkey

1 tsp. salt

2 Tbsp. all-purpose flour

1 Tbsp. Gravy Master (optional)

***Yield*: 4 cups**

It is hard to imagine a Thanksgiving dinner without friends, relatives, and Christine's rich brown gravy. My wife, Christine, has made her gravy the same way for decades, and in all those years, I have never tasted a more delicious rich brown gravy that so beautifully complements the roasted turkey, stuffing, and mashed potatoes. It also elevates any leftovers to a new level, so make enough to have some in reserve.

One of the secrets to the delicious flavor is the homemade turkey stock, building on the rich savory gravy-enhancing flavor of finely chopped onions simmered for several hours. Any stock recipe that uses a whole or quartered onion and doesn't call for simmering the stock for at least several hours (2–4 hours is best) just won't be nearly as good. The stock is then thickened with a roux made of flour and fat in which the starch from the flour gelatinizes with heat, thickens, and stabilizes the fat and water emulsion until it is velvety smooth.

Finely chopping the onion damages countless cells, causing the release of an enzyme called alliinase that reacts with a natural compound in onions called isoalliin to rapidly produce propanethial-S-oxide (PSO), the tear-causing compound. When the finely chopped onion is cooked in water for several hours, the prolonged gentle heating slowly transforms the PSO into a new compound, 3-mercapto-2-methylpentan-1-ol (MMP), which is water soluble. Of the nearly fifty flavor compounds identified in brown gravy produced from beef, pork, or vegetables and finely chopped onions, this simple compound has been shown to have the greatest impact on the savory, meaty flavor of gravy even though it is formed in only minuscule amounts. The more finely chopped the onion, the more PSO and MMP produced. Leaving the onion whole or only cutting through it (as when halving or quartering) produces very little PSO. Your eyes won't tear up, but neither will you produce the compound responsible for the intense, savory flavor of rich brown gravy.

The same process can be used to make gravy for roasted chicken, beef, or pork or vegetable-based gravy. For roasted chicken, use the chicken parts included in the bag found in the cavity of the chicken to make the gravy. Scraps of beef or pork with bones can be used to make beef or pork gravy, and a vegetarian version of gravy can be made with umami-rich ingredients such as mushrooms in place of the poultry or meat in the stock and vegetable oil in place of the animal fat in the roux.

DIRECTIONS:

The total preparation time is about 3 hours, performed while the turkey roasts and rests.

To make the homemade turkey stock, place 4 cups of water in a 2-quart saucepan, and add the onion, carrot, and celery. Submerge the turkey neck and the giblets in the liquid, and add the salt. Bring the water to a slow simmer, and continue to cook uncovered for at least 2 hours, adding water as needed to maintain a fairly constant level in the pan. Strain the hot stock into a large bowl or 1-quart measuring cup, discard the solids, and reserve the stock until the turkey is removed from the oven.

While the turkey rests, pour the drippings from the roasting pan into a fat separator. Make sure to leave any fond (pan scrapings and brown bits) in the pan. Return the defatted drippings to the pan, and reserve 2–3 tablespoons of the fat for making a roux. Add the strained stock to the pan, bring the mixture to a gentle simmer, and scrape up all the fond to distribute it thoroughly into the stock. Meanwhile, make a roux in a small saucepan by thoroughly mixing about 2 tablespoons of the reserved turkey fat with the flour and heating the mixture until it is light golden brown in color. Whisk the roux into the simmering drippings-stock mixture until the roux is evenly dispersed and there are no visible lumps. Slowly simmer the gravy for about 30–45 minutes, until it is velvety and thick enough to coat the back of a spoon.

If you like darker brown gravy with notes of caramel and umami, slowly add Gravy Master a little at a time (be careful not to add too much—a little goes a long way) until the desired color is reached. It also adds to the flavor if you have limited drippings with too little fat (which depends on the size of the turkey and the temperature of the oven). Add salt and pepper to taste.

In the meantime, on August 1, 1774, Joseph Priestley (1733–1804) first prepared oxygen by heating mercuric oxide in his laboratory in Birmingham, England. He also had no idea that he had prepared oxygen; he knew only that it was a component of air that caused a candle to burn brightly with a vigorous flame and in which a mouse could thrive. This latter observation convinced Priestley that the "special air" he had prepared was probably a component of "common air" that was essential for life. He quickly published his observations in three volumes titled *Experiments and Observations on Different Kinds of Air*. The first volume appeared in 1774 and the other two in 1775 and 1777. Priestly is thus given credit for being the first to publish on the preparation of oxygen, while Scheele is recognized as the first to prepare it. Priestley was a self-trained chemist who was a supporter of the French Revolution, a separatist theologian, and a founder of Unitarianism in England. His outspoken views led to arguments and fights with many of the local inhabitants of Birmingham, and after they burned his home and church, he fled to America in 1791, where he became friends with Benjamin Franklin and Thomas Jefferson. Priestley's original laboratory has been reconstructed and is now on display at the Smithsonian Institution in Washington, D.C.

The Scientific Method Illuminates a New Path

In October 1774, shortly after publishing his first volume of *Experiments and Observations on Different Kinds of Air*, Priestley visited Paris, where he met with the brilliant young French chemist Antoine-Laurent Lavoisier (1743–1794) (figure 3.3). When Priestley told him about the "special air" he had produced that caused a candle to burn with a vigorous flame and in which a mouse thrived, Lavoisier was intrigued by these properties but perplexed about how heating the red solid mercuric oxide produced the special air. He was familiar with the earlier work of Boyle, in which the latter performed experiments measuring the precise weight gained when a metal was heated vigorously by a flame in air. Boyle thought the gain in the weight of the metal was due to a combination of the metal with the flame, as fire was considered a substance at the time. The experiments of Scheele and Priestley were qualitative in that the weight of the gas produced by heating mercuric oxide was not measured. Lavoisier had the incredible insight to believe that the increase in the weight of the metal heated in air was caused by a combination of the metal with the special air (he called it "pure air") that had been discovered by Scheele and Priestley. Lavoisier later named the pure air *oxygen* and was also responsible for naming the elements carbon and hydrogen.

FIGURE 3.3

Portrait of Antoine-Laurent Lavoisier and his wife, Marie-Anne Pierrette Paulze, by Jacques-Louis David, 1788, from the collection of the Metropolitan Museum of Art, New York.

To test his hypothesis, Lavoisier devised an experiment for which he is now famous—and that earned him recognition as the father of chemistry. To perform his experiment for the Royal Academy of Sciences just outside Paris, Lavoisier needed a source of intense heat, which turned out to be a device called the "burning glass," a very large magnifying glass lens that could focus the energy of the sun into an intense source of heat. With this source of heat, Lavoisier decomposed solid mercuric oxide into liquid mercury and the pure air (oxygen) discovered by Priestley and Scheele. He carried out the experiment in a closed apparatus (called a *retort*) so the air (oxygen) could be captured and its volume measured precisely, along with the exact weights of the starting mercuric oxide and the liquid mercury product. Then he performed the experiment in reverse and found that the weights of the liquid mercury and pure air (oxygen) consumed, and of the mercuric oxide produced, were exactly the same as when he ran the reaction in the opposite direction, starting with mercuric oxide. He had proved for the first time that the exact weights of a chemical reaction run in either direction are constant and that matter is neither created nor destroyed—the law of conservation of matter. This amazing discovery was published in 1777 in *Memoirs de l'Académie royale des sciences.* Lavoisier went on to describe the invariable laws of nature that govern the process of combustion, publishing his thesis in 1783, also in *Memoirs de l'Académie royale des sciences.* Thus the understanding of combustion as a chemical process of oxidation that generates energy (in the form of heat and light) had progressed from Boyle to Lavoisier, reaching completion in a little more than 100 years.

In his relatively brief life, Lavoisier was interested in many aspects of chemistry, including the process of fermentation, all based on his newfound understanding of the law of conservation of matter. He described fermentation as follows:

> This operation [fermentation] is one of the most striking and extraordinary of all those which chemistry presents to us, and we must examine whence comes the disengaged carbonic gas [carbon dioxide] and the inflammable spirit [alcohol] which is formed and how a sweet body, a vegetable oxide [sugars], can transform itself thus into two different substances, one combustible [alcohol] and the other highly incombustible [carbon dioxide].

Lavoisier continues:

> Upon this principle the whole art of making experiments in chemistry is founded: we must always suppose a true equality between the principles of the body which is examined and those which are obtained on analysis. Thus, since must of grapes gives carbonic acid gas [carbon dioxide] and alcohol, I may say that must of grapes = alcohol + carbonic acid [carbon dioxide dissolved in water produces carbonic acid].

The equation for the conversion of grape must into alcohol and carbon dioxide is the first example of a modern chemical equation. As with Priestley, Papin, and Vavilov, life was not kind to Lavoisier. At the age of twenty-six, he was elected to the Academy of

Sciences and that same year became involved with Ferme générale, a tax-farming (collecting) agency for the royal government. Following the French Revolution, an order was issued in November 1793 to abolish the Ferme générale and arrest all former tax farmers. Lavoisier was convicted of crimes against the people and guillotined on May 8, 1794, at age fifty. In late 1795, the French government informed Lavoisier's widow he was officially exonerated, as he had been "falsely convicted." We can only dream of what great contributions Lavoisier may have made to science had he lived another 25 years. The life of pioneering scientists, especially those who make major contributions to knowledge and social change, has not always been easy.

After nearly 2 million years, fire and combustion were finally understood to be a chemical process of oxidation that generates light and heat, making it possible to explain the true nature of heat, which was first proposed by Count Rumford (1753–1814) in 1798. Until this point, heat was believed to be a weightless fluid called *caloric*, even by such outstanding scientists as Lavoisier. Count Rumford was born Benjamin Thompson in Woburn, Massachusetts. As a young man, he proved to be adept at science, especially mathematics and physics, and occasionally walked the 10 miles from his home to Cambridge to attend lectures at Harvard University. At the age of nineteen, he married a fairly well-to-do, but significantly older, woman (Sarah Walker) and moved to Rumford, New Hampshire, later named Concord and now the state capital of New Hampshire. At the beginning of the Revolutionary War, Thompson favored the British and even did some spying for the British army. When the British evacuated Boston in 1776, Thompson escaped to England, leaving his wife and daughter in America. He became a member of the Royal Society in England in 1779, was knighted in 1784, and moved to Bavaria the following year. There he took a position as a commander in the military and was given the title of count, renaming himself Count Rumford after the town in New Hampshire. One of Count Rumford's duties was to supervise the boring of cannon tubes from solid pieces of metal. Rumford noticed that the friction created by boring the metal produced intense heat, which he assumed was released by the metal as caloric. But then he noticed that the same amount of heat was generated when using a dull drill as when using a sharp new drill, although the dull drill would not penetrate the metal. How could caloric be released if the dull drill could not penetrate the metal? At this moment in 1798, he realized heat could not be a fluid but must be a form of energy generated by the friction created during the drilling process. He actually described heat as mechanical energy, which could be measured as the "mechanical equivalent of heat," later known as the heat capacity or thermal capacity of a substance such a water or iron or olive oil.

Thomas Keller is a great chef who knows a thing or two about cooking. So when he first decided to poach lobster in butter instead of wine or cream, he must have had a good reason (*The French Laundry Cookbook*, 1999). The obvious one is that lobster prepared this way is luscious, moist, and supertender—the perfect marriage of flavor and texture!

Is there any science behind poaching lobster in butter that we can build upon to create new dishes similar to Keller's? The answer is yes, but it may seem counterintuitive. It all has to do with the difference in the *heat capacities* of butter and wine. When heat is applied to a substance, such as butter or wine, the temperature of the substance rises. Mathematically, the quantity of heat added to a substance divided by its rise in temperature is called the heat capacity of that substance (that is, heat capacity equals the quantity of added heat divided by the rise in temperature). The quantity of heat is measured in calories and the temperature rise in degrees Celsius (the temperature scale used in the laboratory). The typical units of heat capacity measured in the laboratory are expressed as calories/gram of substance to raise the temperature by one degree Celsius (°C), or calories/gram (°C). This is exactly equal to 1 Btu/pound (°F) when expressed in the Fahrenheit scale typically used in the United States. For the purposes of this discussion I will explain heat capacity for both the scientific scale expressed in degrees Celsius (formerly centigrade) and Fahrenheit for those not familiar with the scientific scale.

The heat capacities of butter and wine are not the same. It takes more calories of energy to raise wine by 10°C than it does butter. Unfortunately, wine and butter are mixtures with varying compositions, so their heat capacities may vary. But water and olive oil are relatively pure substances, and their heat capacities can be measured very accurately. Using olive oil and water as surrogates for butter and wine, we can explain the difference in their heat capacities. Water has a heat capacity that is about 2.1 times greater than that of olive oil. This means that it takes 2.1 times more calories of heat to raise the temperature of water by 10°C than it does to raise the temperature of olive oil by the same amount.

But why? Recall that temperature is a measure of how fast molecules are moving. Water molecules have a strong affinity for each other due to an electrostatic attraction, which creates relatively weak associations between water molecules known as *hydrogen bonds*. As a result of these hydrogen bonds, it takes lots of energy to pull water molecules away from each other in order to get them to move faster. Molecules of olive oil have much less affinity for each other, so it takes less energy to separate them and make them move faster.

Okay, so it takes about twice as much energy to heat water (or wine) to 160 degrees Fahrenheit (the temperature scale used in the U.S. kitchen) as it does to heat olive oil (or butter) to 160°F (71°C). But what does this have to do with poaching lobster in butter? All that extra energy added to water heated to 160°F is transferred to food cooked in the water or wine. Only half the amount of heat energy is available to transfer to food cooked in olive oil or butter heated to 160°F, so the food cooks more gently and slowly.

This may seem counterintuitive. Ask most people, and they will probably say that *oil is hotter than water*, so food cannot cook more gently in olive oil than water. They remember that oil can be heated to frying temperatures, usually 350°F–375°F (177°C–191°C), while water can't be heated any higher than its boiling point of 212°F (100°C), so they assume oil must be hotter than water! But if we cook lobster in oil and water, *both heated to 160°F*, the water will contain twice as much energy to convey to the lobster as the oil does. The key is the different amount of energy available to transfer to the food when both oil and water are heated to the same temperature.

Not sure you believe this? Here's a simple experiment you can do to convince the nonbelievers. Heat two small pans, one with water and the other with oil, to 135°F (57°C) (no higher!). Now, stick the index finger of each hand into each liquid at the same time. Which one feels hotter? These experiments were done at America's Test Kitchen and published in *Cook's Illustrated* (March & April 2012: 30), where you can read about the results. You may be surprised to find that the water feels much hotter than the olive oil because the water contains more heat energy. At exactly the same temperature, water contains twice as much heat energy as olive oil (for more on heat, temperature and heat capacity read the sidebar on "What is Temperature and How does it differ from Heat" in chapter 2) .

With knowledge of the heat capacity of substances like olive oil, it is possible to cook foods very gently in these substances using the sous vide method of controlled-temperature cooking. Is it any wonder that Keller, after creating his luscious recipe for butter-poached lobster, went on to write an entire cookbook about sous vide cooking, *Under Pressure: Cooking Sous Vide* (Artisan, 2008)? Now you know the rest of the story (Who said that?).

REFERENCE

Resnick, R., and D. Halliday. "Heat and the First Law of Thermodynamics." Chap. 22 in *Physics for Students of Science and Engineering*, Part 1, 466–488. New York: Wiley, 1962.

Later that same year, Count Rumford returned to England to spend more time researching heat. In 1799, he visited Paris where he met Marie-Anne Lavoisier, the widow of Antoine-Laurent Lavoisier, and they married in 1804 (Rumford's first wife had died earlier in the United States). Marie-Anne had assisted her first husband in his laboratory and was thus able to assist Rumford with his research on heat, but other than that, it was an unpleasant marriage. Rumford invented not only the Rumford fireplace, which he designed to radiate more heat, but also the modern kitchen range, the drip coffeepot, and the double boiler. He is also considered to have created the low-temperature method of sous vide cooking in 1799, which he demonstrated by cooking a shoulder of mutton overnight in a machine he had developed for drying potatoes. His method was later described in an essay in which the mutton was referred to as perfectly cooked, extremely tender, juicy, and very flavorful. What we might think of as a new method of cooking is really quite old. Thus Rumford applied his knowledge of heat to produce many useful inventions for the kitchen and cooking. Although he was very gifted and influential, with a history of many significant accomplishments as a scientist, soldier, and statesman, only a handful of people attended his funeral in late August 1814 in the village of Auteuil, less than 4 miles from the center of Paris.

During the mid- to late 1700s, similar developments in science and cooking were occurring halfway around the world in China. As in Europe, a stark dichotomy existed between a relatively small minority of those in the very wealthy and ruling classes and the teeming masses of the poor and less fortunate. During the lengthy Qing Dynasty (1644–1911), the population of China grew from about 150 million in the mid-1600s to 450 million by the mid-1800s. With far too little food to feed the masses, widespread famine spread throughout the country. To counteract the great disparity between rich and poor, the ruling policy makers encouraged the planting of high-starch foods that were not indigenous to China, such as sweet potatoes, regular potatoes, and corn, which soon became a favorite among the Chinese peasants.

For the wealthy and ruling classes, food and cooking were as highly regarded as art and literature, and talented chefs were respected and well paid. As in Europe, the best chefs worked within the imperial palace while another hierarchy of respected chefs worked in private homes as personal chefs for the wealthy and elite. Many of the chefs in private homes were women. In China, there was a third category of chefs who worked in the commercial kitchens of restaurants serving businesspeople and of catering firms supplying the numerous banquets. Unlike in Europe, restaurants have been a significant part of

FIGURE 3.4

Scene on the Yangtze River near Nanjing, China, illustrating the trading of food. Oil on wood panel painted by the author (1988), based on an original sketch by Thomas Allom, 1843.

the business of food and gastronomy in China since about the twelfth century. In fact, the restaurant claimed to be the world's oldest, Ma Yu Ching's Bucket Chicken House, was established in Kaifeng, China, in 1153 CE. The restaurant's signature dish, Chinese "bucket chicken," bears no relation to KFC's "bucket of chicken." Those who could afford to eat out did so on a regular basis, purchasing food from street vendors, tearooms, and both small and very large restaurants. Chefs like Zhang Dong'guan, who worked in the imperial palace of Emperor Qianlong (1736–1795) and favored Suzhou cuisine, were as sought after and as famous as today's celebrity chefs. Poet, artist, and gastronomic food critic Yuan Mei (1716–1798), called the Brillat-Savarin of China, wrote a well-known book on cooking with the title *Suiyuan Shidan* (*Recipes from Sui Garden*), published in 1796.

Unlike the elevated status of gastronomy in China during the 1700s, the state of science was not as advanced or respected, with much of it imported from Europe by Jesuit and Protestant missionaries who focused primarily on astronomy and mathematics. Europe was ahead of China in the sciences of chemistry (except for gunpowder) and physics and in the technology of basic machines, industrial production, and the mechanization of agriculture, while China had developed expertise in the manufacture of silk and porcelain and the large-scale production of tea that was sought after by the Europeans. There is no evidence that China discovered or developed any basic science that was important for cooking during this period, perhaps because cooking was clearly viewed as an art of the highest order rather than a science. The same was true for China's neighbor India, although the styles of cooking, and especially the uses of spices and seasonings, were very different between the two countries. Yet both developed a love for cooking with hot chili peppers, which were introduced to coastal China in about 1670 and found their way to Sichuan in southwestern China in 1749, long after being domesticated in Mexico about 6,500 years ago.

4 The Art of Cooking Embraces the Science of Atoms (1800–1900)

The Atomic Theory Changes Science Forever

In 1799, while Count Rumford experimented with low-temperature cooking, Joseph-Louis Proust (1754–1826), a French chemist, proved experimentally that the elements that comprise chemical compounds always combine in constant proportions. Proust did much of his research with copper carbonate and showed that its proportions of copper, carbon, and oxygen are always the same regardless of how or when the compound is prepared. This became known as the law of constant proportions. This was just the evidence that John Dalton (1766–1844) needed to prove his concept of atomic theory in 1805. Dalton's theory would change science forever. Finally, all matter, everything that exists on earth and beyond, could be understood in terms of its most fundamental building blocks, atoms.

Dalton spent his career as a meteorologist working in Manchester, England, where he started thinking about the physical properties of the gases in air. Out of this evolved his concept of atomic theory, which stated that all chemical elements are composed of extremely small particles of matter that he called atoms (after the Greek word *atomos*, meaning "indivisible"). All the atoms that comprise a single element, such as oxygen, are identical in their weight (called *atomic weight*) and chemical properties, but the atoms in different elements, such as carbon and sulfur, have weights and properties that are different from those of oxygen atoms. The constancy of weight and chemical properties for all the atoms in each element determined the proportion in which they combined with the atoms

FIGURE 4.1

Engraving of John Dalton, by William Henry Worthington, 1814. From the Science and Society Picture Library collection of the Science Museum, London. Reproduced with permission.

of other elements to form chemical compounds such as copper carbonate. For example, Dalton's theory explained why one atom of carbon (C) would always combine with two atoms of oxygen (O) to form carbon dioxide (CO_2), while one atom of carbon (C) would also combine with four atoms of hydrogen (H) to form methane (CH_4), which was known as marsh gas in his day.

With Dalton's theory, it was now possible to understand the structure of chemical compounds based on the particles of matter called atoms and to predict how the atoms of one element would combine with the atoms of other elements to form new chemical compounds. It would soon be possible to predict how the molecules in food, such as water, sugars, fats, and proteins, would behave when heated. Understanding why meat develops such a wonderful flavor when it is roasted and why the starch in potatoes becomes soft and easy to digest when baked in the oven and why vegetables become tender when boiled in water has become possible based on Dalton's atomic theory. As these secrets of cooking are gradually revealed through science, chefs are able to create innovative new dishes with delicious flavors and sumptuous textures. Jacob Bronowski eloquently described Dalton's revolutionary contribution to science in his book *The Ascent of Man*:

> Dalton was a man of regular habits. For fifty-seven years he walked out of Manchester every day; he measured the rainfall, the temperature—a singularly monotonous enterprise in this climate. Of all that mass of data, nothing whatever came. But of the one searching, almost childlike question about the weights that enter the construction of these simple molecules—out of that came modern atomic theory. That is the essence of science: ask an impertinent question, and you are on the way to the pertinent answer.

The meaning of William Blake's poetic lines "Hold Infinity in the palm of your hand, And Eternity in an hour" finally becomes clear when understood in terms of Dalton's atomic theory (the opening four lines of poetry are quoted in the preface to this book). Atoms are so infinitesimally small that, until very recently, we have not been able to visualize them—so small we can easily hold a nearly infinite number in our hand (the number of oxygen and hydrogen atoms in a thimbleful of water is about 2 followed by twenty-four zeros—an extremely large number!). And because the atoms that make up matter can be neither created nor destroyed, according to the law of the conservation of matter, they exist for all eternity.

NUMBERS BOTH LARGE AND SMALL

Numbers can be very confusing, especially if they are not the typical numbers you are used to dealing with, such as the price of a pound of salmon or the balance in your checking account. To make matters more difficult, scientists and much of the developed world follow the metric system, based on units of 10, to measure everything from weights (for example, a kilogram) to distance (for example, a meter), while Americans are saddled with unwieldy units such as 12 inches in a foot and 16 ounces in a pound. The metric system makes it a lot easier to work with very large or very small numbers.

Everyone knows the speed of light is extremely fast. Traveling at the rate of 3.00×10^8 meters per second, in 1 year (called a *light year*), light travels 9.46×10^{15} meters, or almost 1×10^{16} meters. When we see numbers like these, measuring distances in units of 10 is very convenient because 10 multiplied by 10 $(10 \times 10) = 100$, which is designated 10^2. The 2 is called an exponent. Ten multiplied by itself three times $(10 \times 10 \times 10) = 1{,}000$, written as 10^3. The exponents tell us how many times 10 is multiplied by 10, which in turn tells us how large a number is in thousands (10^3), or millions (10^6), or even trillions (10^{12}). The distance light travels in 1 year is approximately 1 followed by sixteen zeros (10 multiplied by 10 sixteen times), or 10,000,000,000,000,000 meters. If a decimal point is placed immediately to the right of the 1 in this very large number, then there are sixteen zeros after the decimal point (count them), equaling 1×10^{16}, the number of meters traveled by light in 1 year. In 1 million light years, light would travel $10^{16} \times 10^6 = 10^{22}$ meters because when multiplying these numbers, the exponents are simply added $(16 + 6 = 22)$. The metric system makes it relatively easy to deal with very large numbers.

Another extremely large number is the number of atoms or molecules in the amount of a substance that equals its atomic or molecular weight in grams. Molecules are substances composed of more than one atom. For example, one molecule of water is composed of two hydrogen atoms chemically bonded to one oxygen atom (H-O-H, also written as H_2O). The atomic weight of hydrogen is 1, and the atomic weight of oxygen is 16, so the molecular weight of a single molecule of water is $2 \times 1 + 16 = 18$. Therefore, 1 molecular weight of water weighs 18 grams, which chemists decided in 1898 should be called a mole (short for molecular weight). If you have ever taken a chemistry class, then you are familiar with the term *mole*. As a point of reference, 1 tablespoon of water weights 15 grams, so 18 grams, or 1 mole of water, is about 1.2 tablespoons of water. Many years earlier (about 1811), an Italian scientist named Amedeo Avogadro (1776–1866) determined that 1 mole of a substance contains 6.022×10^{23} molecules, or approximately 6 followed by twenty-three zeros. This is an extremely large number of molecules in only 18 grams of water. The number of atoms or molecules in 1 mole of any substance is 6.022×10^{23}, which became known as Avogadro's number. As another example, the molecular weight of carbon dioxide (CO_2) is 44 ($2 \times 16 + 1 \times 12$, the atomic weight of carbon), so 44 grams of carbon dioxide equal 1 mole, which also contains 6.022×10^{23} molecules of carbon dioxide. At standard atmospheric pressure and room temperature, 1 mole of a gas such as carbon dioxide fills 22.4 liters (roughly 5.5 gallons). In other words, 22.4 liters of carbon dioxide weigh 44 grams. Similarly, 1 mole of oxygen (O_2), with a molecular weight of 32 (2×16), will fill 22.4 liters and weigh 32 grams, while 1 mole of nitrogen (N_2) will occupy 22.4 liters and weigh 28 grams (2×14).

With the knowledge that 1 mole of each of the gases that make up air (which is about 78 percent nitrogen, 22 percent oxygen, and only 0.04 percent carbon dioxide) occupies 22.4 liters and contains 6.02×10^{23} molecules of that gas, it becomes possible to calculate that 1 liter of air contains about 25×10^{21} molecules, which is how Enrico Fermi calculated the number of molecules in 1 liter of air for his physics class. Further, by calculating the approximate volume of air surrounding the planet and the number of molecules in this volume of air and then by determining the ratio of the two numbers, he could estimate that there was a "probability of one" that each liter of air breathed in by a person living today will contain at least one molecule of air breathed out by Julius Caesar in his dying breath. Of course, Fermi made lots of assumptions to simplify the calculation of this probability, including that all the molecules of air breathed out by Caesar are still in the air rather then being incorporated into trees or humans or anything not in the air.

This brings us to extremely small numbers such as the size of atoms and molecules. If 1.2 tablespoons of water (1 mole) contain about 6×10^{23} molecules, then each molecule has to be incredibly small. If we could take 10 million atoms (for example, atoms of carbon) and line them up one next to the other, the line of atoms would be only about 1 millimeter long (1 millimeter is about 0.04 inches). This would be similar to the point of a pencil containing 10 million (10×10^6) atoms. The diameter of atoms was first estimated by the Swedish physicist Anders Jonas Ångström (1814–1874), who adopted his last name as the unit of measure for the diameter of an atom. Today we measure such small distances in *nanometers* rather than angstroms, but the diameter of an atom is still the same: about 0.1 nanometers, or about 1×10^{-10} meters. One nanometer equals 1×10^{-9} meters, or 0.000000001 meters—a very, very small number indeed.

When we deal with very small numbers in the metric system, we use negative exponents of 10 rather than positive exponents. Just as positive exponents tell us how many times 10 is *multiplied* by 10, negative exponents express the number of times 10 is *divided* by 10. So 10^{-3} means 10 divided by 10 three times, which equals 0.001. Notice that the decimal point is moved three places to the left of the number 1 to form 0.001, so the exponent is -3. Similarly, 10^{-6} equals 0.000001, which is the same as one part per million. Recall that humans can smell some molecules at less than one part per trillion, which is equivalent to 1 molecule dispersed in 1,000,000,000,000 molecules, such as water or air; this also equals one part per 10^{12} parts or about 1 second in 32,000 years! The human olfactory system is exquisitely sensitive. As an example, if you are a nonsmoker, have you ever driven past a car with its window open, with someone smoking inside the car, and smelled the smoke? How many molecules of smoke do you think were in the air you breathed in? Small weights are measured in grams, milligrams (one-thousandth of a gram, or 10^{-3} grams), micrograms (one-millionth of a gram, or 10^{-6} grams), nanograms (one-billionth of a gram, or 10^{-9} grams), and pictograms (one-trillionth of a gram, or 10^{-12} grams). Do you think there may have been 1 nanogram of smoke in the air you breathed—or perhaps even less?

REFERENCE

Morrison, P., P. Morrison, and the Office of Charles and Ray Eames. *Powers of Ten: A Book About the Relative Size of Things in the Universe and the Effect of Adding Another Zero*. New York: Scientific American Library, 1982.

Just imagine that the atoms in the carbon dioxide breathed out by your great-grandparents who are no longer physically with us are still present in the atmosphere, as well as in organic matter still present today on the planet. Even more astounding is the famous proposition known as Caesar's Last Breath, posed by Enrico Fermi (1901–1954), Nobel laureate in physics. He calculated that there is a "probability of one" that every breath of air you take contains at least one molecule of air (comprised of oxygen, nitrogen, and a small amount of carbon dioxide) breathed out by Julius Caesar in his last dying breath. Each breath of air (about 1 liter) contains about 25×10^{21} molecules (equal to 25 followed by twenty-one zeros—another extremely large number), making it almost certain that at least one of those molecules you just breathed in was in Caesar's last breath. Statistically, a probability close to 1 means there is a strong chance an event will occur, while a probability of 1 means an event will almost certainly occur. In this sense, we really never lose contact with our late loved ones. The great Indian-born American author Deepak Chopra, expressed this so beautifully when he stated, "The fragrance of your ancestors lingers here right now." A truly powerful thought.

Science Enters the Realm of Cooking

The artistic creativity of cooking finally fused with the new science of Dalton, Antoine-Laurent Lavoisier, and Robert Boyle following the French Revolution. Of all creative activities, cooking best embodies the elements of art and science. I suspect that's why I find cooking so appealing and satisfying. Perhaps no other chef of this era combined art and science in his cooking more than the "King of Chefs," Marie-Antoine Carême (1783–1833). The great chef Auguste Escoffier once said, "The fundamental principles of the science [of cooking], which we owe to Carême, will last as long as cooking itself," while Carême himself stated, "The chef committed to science is more responsive to the praise given by his Patron than to the handful of gold that he might receive from him." Clearly, Carême was a true artist, as seen in the magnificent pastries and dishes he created, but much of his success was due to the science applied in making his creations.

Abandoned by his parents in the streets of Paris at the age of eleven, Carême did not have an easy start in life. Fortunately, by the age of sixteen, he had found his way to an apprenticeship at a Parisian pastry shop and was discovered by his first patron, Charles Maurice de Talleyrand-Perigord (known simply as Talleyrand), the minister of foreign affairs during Napoleon's reign. At that time, the best chefs in France worked for wealthy patrons rather than in commercial restaurants, as food shortages following the French Revolution limited the availability of fresh ingredients and most people could not afford to eat in restaurants. After serving Talleyrand for many years, Carême went on to work for a number of other influential patrons, including the tsar of Russia and Baron de Rothschild. During this time, he wrote a number of major books on French cooking, claiming he reached a far greater audience through writing than all the meals he cooked as a private chef. His reputation soon spread, establishing him as the most celebrated chef in Europe, the creator of nouvelle cuisine, or modern French cooking.

It was no coincidence that French cooking reached its zenith with Carême's nouvelle cuisine not long after French science achieved similar status with Lavoisier's exquisite experiments. Carême's beautiful cuisine was based on science. Despite what you might have read about his elaborate dinners, he believed in maintaining simplicity; using seasonal ingredients; reducing the number of spices (typically parsley, tarragon, and chervil); simplifying sauces, often made from a small number of concentrates; and serving meats in their natural juices rather than heavy gravies. In his final book, *L'Art de la cuisine française* (1833), he listed recipes for almost 300 soups and 358 sauces. It was clear from his use of ingredients such as anchovy butter and shrimp paste made from the tails of shrimp that he was aware of the savory taste of umami well before it was identified almost one hundred years later, in 1908, by a Japanese physical chemist. Once Carême had gained a reputation as a celebrated chef, he was able to live a very comfortable life, which unfortunately came to an early end at the age of forty-nine, presumably due to tuberculosis.

During the French Revolution, Napoleon Bonaparte (1769–1821) developed into a prominent statesman and military leader, ultimately becoming the emperor of France from 1804 to 1814 during the Napoleonic Wars (1799–1815). He recognized very early that an army fights on its stomach and must be constantly supplied with safe, nutritious food. Unfortunately, as his armies conquered ever more territory, it became increasingly difficult to preserve food transported over long distances for extended periods of time. In 1795, at Napoleon's request, the temporary French government, known as the Directory, offered a sizable reward of 12,000 francs to anyone who could develop a new method to prevent food from spoiling. Nicolas Appert (1750–1841), a brewer, confectioner, and chef, answered the call and eventually won the coveted prize in 1810.

Appert was born in Châlons-sur-Marne (renamed Châlons-en-Champagne in 1998), located about 92 miles (148 kilometers) from Paris in the heart of the Champagne region in northeastern France. In 1780, he moved to Paris and opened a small but successful confectionary shop. From a very early age, he experimented with food and acquired an extensive knowledge of food preparation and the methods available at the time for preserving foods. These methods could be divided into two categories: one involved desiccating the food, which is how salting and smoking meat and fish work, and the other called for adding foreign substances such as sugar or vinegar (pickling) to prevent fermentation and putrefaction. Both methods had undesirable effects on the texture and flavor of the food. To make matters worse, no one knew why these methods worked, which may explain why they were not very successful in preserving foods for very long periods. The general theory at the time was that exposure to air causes food to spoil. It would be another 60 years before the culprit was discovered by Louis Pasteur to be nearly invisible microorganisms (frequently coming from the air).

In his experience working with food, Appert had learned that "heat has the essential quality to arrest decomposition" and that "heat applied in the proper manner effects perfect preservation after having deprived them [the food] in the most rigorous manner possible of contact with air." On the basis of this knowledge, Appert set out to devise a new method for the long-term preservation of food. For more than 10 years, he carried out extensive tests to perfect his new method, which involved both heat and rigorous exclusion of air. His process required four steps: (1) place the food in wide-mouthed glass bottles, (2) cork the bottles with great care "because success depends chiefly on the closing," (3) submit the enclosed food to boiling water (in a water bath) for more or less time according the nature of the food, and (4) remove the bottles from the water bath at the prescribed time. His theory was that heating the tightly sealed bottles (the corks were inserted with pressure, sealed with "fish glue," and secured with wire) in boiling water drove out all traces of air from inside the food, thus preventing putrefaction (figure 4.2). His method was fortuitous for several reasons. First, he chose to use glass bottles because he knew glass was impermeable to air (presumably from his knowledge

FIGURE 4.2

The glass bottle used by Nicolas Appert (ca. 1809) and one of the first "tin" cans developed by Peter Durand (1810). Watercolor on paper by the author, based on early photographs.

of bottling champagne), and, second, this choice meant he had to use relatively small bottles, as large glass bottles are more easily broken. With small bottles, he could heat the food completely through, raising even the center of it to the same temperature as boiling water. Little did he know that this process killed all the harmful microorganisms that were hiding in the food. If he had used much larger containers, he might not have heated the food completely through and might not have destroyed all the harmful microorganisms. As we shall see, this later became a fatal problem in the hands of a less scrupulous food processor.

During his lengthy trials, Appert prepared dozens of different preserved foods, including roasted meats, poultry, fish, stews, soups, sauces, vegetables, fruits, and desserts. The meat dishes were all partially cooked before they were placed in the jars and heated, which typically required between 1 and 2 hours in the boiling water bath. The length of time was arrived at by trial and error. In 1804, Appert opened an experimental facility for the large-scale preservation of food, using his new method to produce enough food for testing by the French navy. Following this successful trial, he presented his results to the government in 1809, which approved his method and requested that he write a complete report on his process for preserving foods before awarding him the prize money in 1810, approximately 14 years after he began his work. His report on *The Art of Preserving All Kinds of Animal and Vegetable Substances for Several Years* was 107 pages long and included a detailed description of his method of bottling the food, followed by 61 pages of recipes (Appert 1812). With the prize money, Appert set up a company and small factory for preserving food in 1812; it continued in operation until 1933.

Appert's method drew lots of attention, especially from Peter Durand, a British inventor, who determined Appert's method would be better suited for nonbreakable metal containers, a new invention at the time. Durand obtained a patent for his method in 1810, using soldered iron containers lined with tin to better resist acidic foods. An entirely new method of preserving food called *canning* had been born— and, with it, what is now known as the tin can (figure 4.2). In 1812, two Englishmen, Bryan Donkin and John Hall, purchased Durand's patents; formed the company Donkin, Hall, and Gamble; and turned the invention of canning into a commercial success, of sorts. The method was very expensive, and the cans had to be opened with a hammer and chisel (the can opener would not be invented until 1855). The major customers for canned foods were the governments of Britain and France, which were thus able to supply their armies and navies with safe, nutritious, good-tasting, nonperishable food. As far as satisfying Napoleon's original request to provide a method of preserving food for the military, canning was a great success.

During the early 1800s, the introduction of canned food proceeded relatively slowly, primarily because the cost made it unaffordable to all but the wealthy and the government. Producing the tin-lined cans alone required more than a dozen steps, as the seams had to be soldered with lead inside and out and then along the outside at the top and bottom, once the can was closed. But those who were fortunate enough to buy canned food from the primary manufacturer in England— Donkin, Hall, and Gamble—raved over the quality and the flavor of the food. And there were no complaints about its safety or spoilage. Britain's Royal Navy quickly became the major customer, buying large quantities to supply its ships, especially those going on long voyages and extended expeditions. In 1845, the Royal Navy embarked on the largest, most expensive expedition yet undertaken in an effort to find the lucrative Northwest Passage. This would involve two naval ships, the HMS *Erebus* and HMS *Terror*, which had to be specially outfitted for use in Arctic waters and stocked with enough provisions to feed 129 men for 3 years. The expedition was under the command of Sir John Franklin, who led a handpicked crew of officers and sailors (plus a dog named Neptune and a pet monkey named Jacko). The Royal Navy determined the expedition would require 29,500 cans of dozens of different varieties of precooked canned meats, vegetables, soups, and sauces, including more than 20,000 cans of soup (each weighing 1 pound) and 9,500 cans of meat and vegetables (ranging in weight from 1 to 8 pounds each). The expedition was scheduled to sail on May 19, 1845, yet the order for all the provisions was not issued by the navy until April 1, 1845, making the task nearly impossible to complete on time and within budget. To make matters much worse, the Royal Navy chose to select the winning bid based on the lowest cost! Reputable suppliers such as Donkin, Hall, and Gamble quickly dropped out of the competition, which was won by Stephan Goldner, owner of a small, unknown, unsanitary canning plant in the squalid section of London's East End.

Goldner may have won the contract because he had patented a process for heating canned food to a temperature of 250°F (121°C) using a concentrated solution of calcium chloride (which boils at a much higher temperature than pure water) rather than a bath of boiling tap water. Goldner believed he could use his patented process to can food faster than other companies and therefore meet the deadline while keeping costs low. His was by far the lowest bid of those submitted. But Goldner was an unscrupulous person willing to do whatever it took to make a profit at the expense of the Royal Navy. He used extremely cheap cuts of old meat loaded with heavy bones and spoiled or wilted vegetables, and he hired dozens of unskilled, unsanitary laborers off the street who took shortcuts in producing the cans and processing the food. Apparently, the navy never inspected Goldner's blood-splattered plant, or it might have pulled the plug on the contract. By the end of the first month of the contract, Goldner had not delivered a single can of food! The navy was beginning to worry (panic is more likely), so Goldner asked to package the food in larger cans, up to 12 pounds each, which he argued would allow him to produce the food faster, and the navy agreed to this without objection. In fact, most of the food was packaged in cans that were much larger than the 1-pound cans originally specified by the navy. Finally, all of the cans were delivered to the docks just 48 hours before sailing. Naturally, in its haste to meet the sailing deadline, the navy never checked the contents of the cans and simply accepted the food without objection and loaded all of it below decks on the ships. Goldner had met the deadline and received full payment for his effort. He was never heard from again after 1852. Much of this information is detailed in the original government records of the time.

The fate of Franklin's expedition is now well known, with all 129 members lost; they either froze or perished from some disease like pneumonia or from an unknown cause—perhaps food poisoning or even cannibalism. Numerous rescue attempts and searches have been undertaken over the years. It became apparent that the ships had been completely ice bound for more than 2 years and that the crew eventually abandoned them to search for an escape route over the ice and snow. In 1984, several well-preserved frozen corpses of the expedition's crewmen were found and examined by Dr. Owen Beattie of the University of Alberta. In 2014 and 2016, the two ships were found sunken but in remarkably good condition.

Some of the corpses showed very high levels of lead in their tissue, and until recently, most investigators believed the crew may have been weakened by and succumbed to lead poisoning from the lead pipes used to transport drinking and cooking water inside the ships and from the lead solder used to seal the cans of food. But evidence is now mounting that many of the deaths might be linked to improperly canned food contaminated with botulinum, a fatal neurotoxin produced by *Clostridium botulinum* bacteria. One corpse was found to contain a species of *Clostridium* spores in its intestine. The food poisoning may have been caused by the large cans Goldner used in his rush to complete the order. Completely heating the food in a 12-pound can, even in a 250°F (121°C) calcium chloride water bath, takes far longer than heating food in a 1-pound can (the exact time depends on the food), potentially resulting in cold spots in the middle of the food where *Clostridium* spores would not be killed. *Clostridium* spores must be heated to 240°F–250°F (115°C–121°C) for 3 minutes to be completely destroyed. Yet Goldner heated the 12-pound cans for the same length of

(CONTINUED ON PAGE 78)

RECIPE 4.1 Baked Haddock with Special Breadcrumb Topping

INGREDIENTS:

1 Tbsp. extra virgin olive oil

1 ounce red bell pepper, finely chopped

1 garlic clove, finely chopped

¼ cup dry vermouth, plus additional for the fish

2/3 cup freshly ground dried bread, preferably French-style baguette

1 Tbsp. butter

12- to 13-ounce wild-caught haddock fillet, skin removed

Fresh-squeezed lemon juice

Chopped parsley

NOTE:

This recipe can be easily scaled up to almost any amount and stored in portions in the freezer for many months.

Yield: Enough breadcrumbs for 2 servings of fish (see note)

Would you like to eat more fish but are tired of so much salmon and find white fish like haddock, hake, and cod (members of the same family of fish) are too bland and uninteresting for your taste? Here is a recipe for a special breadcrumb topping that makes even bland-tasting white fish really delicious, and it is quick and easy to prepare for weeknight meals, as well as for special company. Many people are reluctant to cook white fish at home because it is so easy to overcook, making it dry and flaky, and it's also pretty expensive, so you don't want it to be disappointing. The secret to preparing great white fish is to bake it with a breadcrumb topping that prevents the fish from drying out and overcooking and is also really flavorful. The oily breadcrumbs keep moisture in the fish and insulate it from overcooking. The topping in this recipe was developed by my father-in-law to use on baked stuffed shrimp in his New England seaside restaurant. I decided to use it on baked white fish to solve the problem of dull, overcooked white fish.

The topping derives its signature flavor from finely chopped red bell pepper and garlic sautéed in oil and finished with a touch of vermouth. Almost any type of breadcrumb will work, but I prefer freshly ground dried leftover French baguette and sometimes ground Ritz crackers, the classic base for a topping (although the baguette provides a better texture and appearance). Packaged panko-style breadcrumbs also work, but don't use the regular packaged breadcrumbs sold in the supermarket, or they will impart a very dull, musty flavor to the topping. Also, read the label on any store-bought prepared breadcrumb topping, as some still contain partially hydrogenated vegetable oil or shortening, a source of harmful trans fat.

For a great healthy meal, serve the fish on top of mashed cauliflower (see recipe in chapter 6), along with leafy greens (spinach, Swiss chard, mustard greens, bok choy, etc.) sautéed with garlic and oil, and carrots roasted with a little brown sugar and butter.

DIRECTIONS:

Preheat the oven to 350°F. In an 8-inch stainless steel skillet, heat the olive oil on medium heat, add the bell pepper, and sauté for about 4 minutes, until the edges of the pepper just begin to darken. Add the garlic, and continue cooking for about 1 minute; then allow the pan to cool off the heat for about 1 minute (to prevent the vermouth from splattering in the hot pan). Add the vermouth, and continue cooking until virtually all of the vermouth has evaporated. Allow the pan to cool off heat for another minute, add the breadcrumbs, and mix well. Season the breadcrumbs with a small amount of salt and freshly ground pepper, if desired.

Place tiny dabs of butter evenly on the bottom of a shallow baking dish, place the fish fillet on top of the butter, and splash with a small amount of additional vermouth. Spread the breadcrumb topping evenly over the fish, and bake on the middle rack of the oven for 20 minutes. Season with a little fresh-squeezed lemon juice and chopped parsley, divide into two portions, and serve.

time as the 1-pound cans because he had no idea that heat was required to kill any pathogenic microorganisms (and because he was rushing to fill the order). *Clostridium botulinum* bacteria were unknown until 1897. In 1845, it was assumed, even by Appert, that heat was required only to drive all the air out of the food. Once sealed inside the can, the anaerobic *Clostridium* spores happily resided there until activated by heat, resulting in the growth of the bacteria and the production of the deadly neurotoxin. With fuel running short, some of the canned food, especially the food taken along by the sled parties looking for an escape route out of the ice fields, may not have been heated at a high enough temperature or for a long enough time to destroy any resident toxin. Destruction of 99 percent of botulinum toxin C requires that food be heated at 176°F (80°C) for 5 minutes.

Aside from this tragic example of taking shortcuts, canning has proved to be a very safe way of preserving food. In fact, in 1974, the National Food Processors Association opened and tested a number of foods that had been canned in 1865. The appearance, smell, and vitamin content had deteriorated, but there were no traces of microbial contamination. The food was judged to be safe to eat, if less than appealing.

The Art of Cooking Embraces the Science of Cooking

During the period from 1820 to 1860, there were great advances in science, especially in food chemistry, in both France and Germany. Competition between the academic scientists in these two countries was especially fierce and yet very constructive. By the late 1830s, they had established that plants and animals contained sugars, fats, and nitrogenous substances soon to be identified as proteins. They found that, together with water, these substances comprised all the macronutrients present in food and therefore all the components involved in the transformations that occur when food is cooked. When I was in graduate school many years ago obtaining my PhD degree in organic chemistry, it was required that every student prove he or she was proficient in reading and writing scientific German before graduating. For those like myself who are not adept at learning languages, scientific German is especially difficult, as some sentences can be an entire paragraph in length, and many words are actually compound words created by putting multiple words together into one long word. For example, the German word for the number 7,254 is *siebentausendzweihundertvierundfünfzig*! The longest word in the *Guinness Book of World Records* is a German word containing thirty-nine letters—*rechtsschutzversicherungsgesellschaften*, which means “insurance companies providing legal protection.”

So why am I writing about the difficulties of German scientific language? Because in 1821 Friedrich Accum (1769–1838), a German chemist, wrote one of the early books on cooking science, and it has perhaps the longest title of any book on science or any other topic: *Culinary Chemistry Exhibiting the Scientific Principles of Cookery with Concise Instructions for Preparing Good and Wholesome Pickles, Vinegar, Conserves, Fruit Jellies, Marmalades, and Various Other Alimentary Substances Employed in Domestic Economy, with Observations on the Chemical Constitution and Nutritive Qualities of Different Kinds of Food*. Fortunately, it is simply referred to as *Culinary Chemistry*. Despite the lengthy title, the book sold very well and helped pull Accum out of poverty. He wrote that "the art of preparing food is a branch of chemistry: The kitchen is a chemical laboratory." Aside from being rather verbose, Accum was very instrumental in bringing to the public's attention the widespread problem of adulterated processed food, which he referred to as "death in the pot." Not to be outdone by the Germans, Jean Brillat-Savarin (1755–1826), the famous French gastronome and amateur food scientist (a lawyer by profession), published his famous book *The Physiology of Taste* in 1825, just two months before his death. The full title of his book easily challenges Accum's: *The Physiology of Taste, or, Meditations of Transcendent Gastronomy; A Theoretical, Historical and Topical Work, Dedicated to the Gastronomes of Paris by a Professor, Member of Several Literary Societies*. I will not attempt to add any more about Brillat-Savarin's life, as he is so well known in cooking and gastronomy circles and the title of his book says it all.

Justus von Liebig (1803–1873), one of the best-known German chemists of the time, made significant contributions to food chemistry and the science of cooking. He was an extremely productive chemist with an extensive body of very influential work in pharmacy, organic chemistry, agricultural chemistry, physiological chemistry, food chemistry, and even the chemistry of sewage. He was also a very careful experimentalist, often having his students recheck and validate the work of other scientists. Among the many works that Baron von Liebig wrote, one of his most important, published in 1847, was given the thankfully brief title *Research on the Chemistry of Food*. A number of scientists from France (Michel Chevreul, Jean Dumas), the Netherlands (Gerardus Mulder), Sweden (Jöns Berzelius), and Germany (Friedrich Wöhler) were involved in identifying the sugars, fats, and proteins in food. However, Liebig made some of the most significant contributions, especially regarding the nutritional importance of these constituents of food to good health. He showed that respiration was a slow form of combustion in which the organic matter of food was oxidized to form the carbon dioxide expelled in the breath and that the heat generated by the reaction was the source of body heat. He went on to say the "the unassimilated nitrogen of the food, along with the unburned or unoxidized carbon is expelled in the urine or in the solid excrements."

One of Liebig's most successful ventures was the development of "meat extracts" as a cheap, nutritional food for the poor, as well as for ill and invalid patients, and in 1862, he formed the Liebig Extract of Meat Company along with a partner, Georg Giebert. With Giebert's business guidance, the company made Liebig a wealthy man. Today the company is part of the multinational company Unilever. Liebig's process involved boiling chopped meat in eight to ten times its weight in water for half an hour to dissolve all of the active nutritional ingredients, remove the fat, and concentrate the extract by evaporation. In this manner, 33 pounds of meat could be reduced to 1 pound of "essence, which is sufficient [when reconstituted] to make broth for 128 men."

Unfortunately, in today's world of cooking lore, Liebig is better known for the error he made rather than for all his accomplishments. He insisted that, when meat is added to boiling water, the protein on the surface coagulates and forms a barrier against the loss of juices and nutrients. Cooks have assumed ever since that the surface of meat seared in a hot pan or oven forms a crust, preventing the loss of juice from the meat. But Liebig never extended his supposition to meat seared with dry heat. Research conducted at the University of Missouri in 1930 proved searing did not prevent loss of juice by showing that roasts cooked in a very hot oven lost more moisture than similar roasts cooked in a colder oven. In fact, the amount of moisture lost from cooked meat has been shown to be directly proportional to the internal temperature of the meat. The myth that braising meat in liquid makes it juicier has been disproved by research conducted in the 1970s and 1980s that showed the heat transferred from a simmering braising liquid such as water causes the muscle fibers to shrink and squeeze out as much moisture as in meat roasted in a dry oven.

There is one other area of cooking in which Liebig played a role that many cooks may not realize. One of his American students, Eben Norton Horsford (1818–1893), became a professor of chemistry at Harvard and obtained a patent in 1856 for a new baking powder that could replace yeast in baked goods. Horsford founded the Rumford Chemical Works in Rumford, Rhode Island, to manufacture the product, which was sold as Rumford Baking Powder. In Germany, Liebig promoted the product as "the chemical method for making bread" from refined white flour and encouraged two of his German students to manufacture the product for Europe, using a mixture of calcium and magnesium phosphates and sodium bicarbonate (baking soda). In 1868, Liebig showed that white bread made from refined white flour (flour from which the bran had been removed) was nutritionally inferior to bread made with whole wheat flour.

DOES BRAISING MEAT REALLY MAKE IT JUICIER?

One of the most common cooking myths is that braising meat makes it juicier. After all, braising is a method of cooking meat in a liquid such as stock or wine. It seems common sense to assume that some of the liquid would make its way into the meat as it cooks and make it juicy. Before deciding whether this is fact or fiction, we need to examine the structure of meat and what makes it juicy or dry when it is cooked by different methods.

Meat is the muscle tissue of an animal. It is composed primarily of muscle fibers, connective tissue, fat, and water. The first two are proteins. The dominant proteins in muscle fibers are *actin* and *myosin*, while the primary protein in connective tissue is *collagen*. A deconstructed picture of muscle and its component parts is shown in figure 1. Muscle is composed of thousands of muscle fibers, each of which is composed of very long individual muscle cells. The muscle fibers are packed into bundles surrounded by connective tissue. Each fiber is composed of hundreds of small tubules called *myofibrils*, which are made up of bands of proteins called *actin* (thin filaments) and *myosin* (thick filaments) that control muscle action. When muscle contracts, the actin and myosin proteins bond together through the formation of chemical cross-links, causing the proteins to move closer together. When the muscle relaxes, the cross-links are broken, and the actin and myosin return to their original positions. Actin and myosin are held in position by other proteins called *Z-disks*.

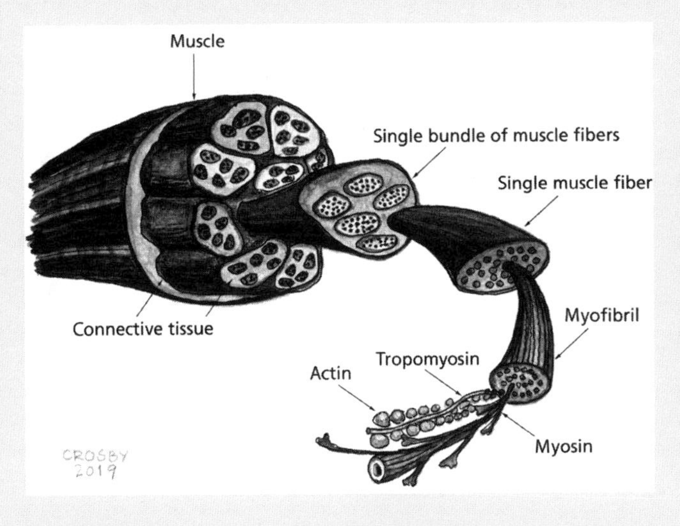

FIGURE 1

The composition and structure of a muscle fiber. Watercolor and ink on paper by the author.

The meat from animals is about 75 percent water by weight. About 80 percent of this water is contained within the myofibrils in the spaces between the thick and thin filaments (see figure 1). When meat is cooked to high temperatures (reaches the well-done stage), the myofibrils shrink in diameter, squeezing out some of the water inside them. Shrinking actually starts at temperatures as low as 104°F (40°C), with maximal water loss beginning at 140°F (60°C). Muscle fibers have been shown to shrink very rapidly to about half of their original volume when heated to temperatures between 122°F (50°C) and 158°F (70°C). If a piece of meat is cooked and immediately sliced, the juice we see running out of the meat is the water that has been squeezed out of the myofibrils into the spaces between the muscle fibers, where it is free to escape as soon as the meat is sliced. Active shrinking of the myofibrils during cooking is the major cause of water loss in meat.

The ability of meat to hold onto the water within muscle tissue is called its *water holding capacity* (WHC), which determines how much water the meat will retain (or lose) when cooked. Related to the WHC of meat is its ability to absorb water from a brine solution. One of the most important factors in determining the WHC of meat is the pH of the meat. Because the juiciness of cooked meat is perceived as tenderness, the pH of raw meat is perhaps the single most important factor in determining the eating quality of cooked meat.

Many of the amino acids that link together like a long chain of paper clips to form proteins are electrically charged. The electrical charges of different regions of a protein greatly affect how it interacts with its neighboring proteins. These charges are directly influenced by the pH of the environment surrounding the protein. In an acid environment, the protein acquires additional positive electrical charges, while in an alkaline environment, it acquires additional negative electrical charges. At a pH called the *isoelectric point*, all of the positive and negative electrical charges of a group of neighboring proteins are completely balanced, so the proteins are electrically neutral. At this pH, there is no repulsion between proteins, so all of the proteins can cluster together like a crowd of people. When muscle proteins pack together this tightly, there is little room for water inside the myofibrils. In addition, tightly packed muscle fibers are difficult to bite through. Thus, when the pH of meat is at its isoelectric point, it will be very tough and dry when cooked.

The isoelectric point of most meat such as beef and pork is about pH 5.2, or mildly acidic. As the pH rises and becomes less acidic, the muscle proteins acquire additional negative electrical charges that force the proteins apart (remember that like charges repel each other), creating more space for water and making the muscle fiber easier to bite through. The quality of lean pork is strongly influenced by the pH of the meat. To be tender and juicy, pork should have a pH of 6.5 or higher. Fortunately, it is relatively easy to judge the pH of pork, as darker pork has a higher pH. Select pork that is relatively dark and well marbled with fat, and cook it to an internal temperature of 145°F (63°C) at the thickest part of the meat, according to the new U.S. Department of Agriculture guidelines. The center should still be pink. Beef tends to be more acidic, with a pH around 5.5–6.0. But since a whole cut of beef is sterile inside, it can be cooked to 125°F–130°F (52°C–54°C) without danger of food poisoning. As we will see, a lower internal temperature results in more tender, juicy meat. Before I finally answer the original question about braising meat and juiciness, let me just add that the pH of meat is largely dependent on how the animal was treated shortly before slaughter. The more stressed the animal is, the more lactic acid that builds up in the muscle tissue following slaughter, thus lowering the pH of the meat. More stress = more acid = lower pH.

So does braising meat make it juicier? The answer is *no*! For meat at a given pH, the amount of water it retains when cooked is directly related to its temperature not to how it was cooked. This is because the temperature of the meat determines how much the muscle fibers shrink, how much water they can hold, and how tender the meat will be. The research supporting this conclusion goes back 40 years. Several research papers published in the *Journal of Food Science*, *Journal of Animal Science*, and *Meat Science* (see the references) in the 1970s and early 1980s support this conclusion. It is surprising the myth that braising meat makes it juicier has lasted this long.

The following table summarizes the results published in the *Journal of Food Science* (McCrae and Paul 1974), in which the researchers compared "cooking losses" in braised and roasted beef steaks cut to a thickness of 1 inch (2.5 centimeters) from the same cut of eye of the round beef. Both braised and roasted meats were cooked to the same internal temperature of 158°F (70°C).

The oven temperature for roasting the meat was 325°F (163°C). Water was used as the braising liquid.

Measure	Braised	Roasted
Total cooking loss	29.58%	28.20%
Drip loss	20.02%	5.62%
Evaporation	9.57%	22.58%

Notice the amount of water lost (along with a small amount of fat) is essentially the same for both dry and wet cooking methods. But, interestingly, the ways the water was lost were different. Most of the water lost during braising was the result of drip loss, while most of the water lost during roasting was the result of evaporation, which seems quite logical. *It is very clear that braising meat in liquid does not make it juicier.*

REFERENCES

Bendal, J. R., and D. J. Restall. "The Cooking of Single Microfibers, Small Microfiber Bundles and Muscle Strips from Beef Muscles at Varying Heating Rates and Temperatures." *Meat Science* 8 (1983): 93–117.

Bengtsson, N. E., B. Jakobsson, and M. Dagerskog. "Cooking of Beef by Oven Roasting: A Study of Heat and Mass Transfer." *Journal of Food Science* 41 (1976): 1047–1053.

Cross, H. R., M. S. Stanfield, and E. J. Koch. "Beef Palatability As Affected by Cooking Rate and Final Internal Temperature." *Journal of Animal Science* 43 (1976): 114–121.

McCrae, S. E., and P. C. Paul. "The Rate of Heating As It Affects the Solubilization of Beef Muscle Collagen." *Journal of Food Science* 39 (1974): 18–21.

Offer, G., and J. Trinick. "On the Mechanism of Water Holding in Meat: The Swelling and Shrinking of Myofibriles." *Meat Science* 8 (1983): 245–281.

Schock, D. R., D. L. Harrison, and L. L. Anderson. "Effect of Dry and Moist Heat Treatments on Selected Beef Quality Factors." *Journal of Food Science* 35 (1970): 195–198.

THE MANY LIVES (AND USES) OF BAKING SODA

Baking soda, aka sodium bicarbonate ($NaHCO_3$), has been used as a leavening agent for baked goods since at least 1869. When heated or mixed with acid, baking soda forms carbon dioxide gas (and water as a by-product of the reaction). It is frequently used alone, but it is more often used in the form of baking powder, in which it is combined with certain acids, such as monocalcium phosphate. It is also common practice to use baking soda and baking powder together when a recipe involves acidic ingredients such as buttermilk. Without the addition of baking soda, the acid in buttermilk would quickly neutralize the baking powder, which is designed to release much of its gas slowly in the oven.

All of this is well known to cooks. But baking soda has many other applications in the kitchen that are less well known. These applications are based on the fact that sodium bicarbonate is a mild alkali. Most food ingredients are neutral or acidic. Egg whites are the only other commonly used alkaline food ingredient.

For many years, baking soda has been used to hasten the cooking of vegetables. It does this because the added alkali greatly accelerates the breakdown of pectin, which strengthens plant cell walls and holds the cells together. Nowadays cooks prefer their vegetables firm, so baking soda is not used very much for this purpose. But there are times when a pinch of alkaline baking soda makes cooking easier. Take the example of "No-Fuss Creamy Polenta," published in *Cook's Illustrated* (March & April 2010: 18–19). The addition of a small amount (literally just a pinch) of baking soda to the water used to cook the coarse-ground cornmeal softens the cell walls of the cornmeal and dramatically reduces both the cooking time and the amount of stirring. It takes just a pinch because once the alkaline baking soda initiates the breakdown of pectin, the reaction proceeds on its own.

In another application, increasing the amount of baking soda in a recipe for "Crisp Gingersnap Cookies" (*Cook's Illustrated*, November & December 2011: 22–23) helps to develop desirable fissures in the cookies, as well as to produce drier cookies with increased browning and flavor. In this case, the baking soda weakens the development of gluten, creating a more porous structure. More-alkaline dough also promotes cookie spread, as well as browning and flavor by the Maillard reaction (for detailed discussion of the Maillard reaction see the section on "Flavor Rules" in chapter 5).

Use of a small amount of baking soda also makes it possible to make "Great Home Fries" (*Cook's Illustrated*, January & February 2012: 12–13). When the potato cubes are cooked in water with a little baking soda for just 1 minute, the exteriors of the potatoes become mushy, releasing the starch molecule known as amylose, which helps to make a crispy brown crust when the potatoes are roasted in the oven. Meanwhile the interiors remain firm and moist, so they cook up creamy and tender. In this case, the baking soda hastened the breakdown of the potato cell walls, enabling the release of amylose.

As a final example, baking soda helps to tenderize meat. The tenderness of pork is dependent on the pH of the meat. Up to a point, the higher the pH of the meat is, the more tender it will be. Consider a recipe for "Sichuan Stir-Fried Pork in Garlic Sauce" (*Cook's Illustrated*, March & April 2012: 10–11). After the pork was soaked for only 15 minutes in a dilute solution of baking soda (made by dissolving 1 teaspoon of baking soda in ½ cup of water), it cooked up to be juicy and supple. Increasing the pH of meat makes it less acidic and helps it to retain more moisture, which helps it to remain tender.

To understand how these reactions occur, let's look more closely at the chemistry of baking soda. When something is alkaline, it means a solution of this substance in water has a pH greater than 7, which is neutral. A solution with a pH below 7 is acidic. The pH scale (0–14) is a measure of the concentration of hydrogen ions (H^+) and hydroxide ions (^-OH) in solution. When dissolved in water, acids produce an excess of hydrogen ions, which combine with water to form hydronium ions (H_3O^+), while alkalis produce an excess of hydroxide ions. At neutral pH 7, the two ion concentrations are equal. A 5 percent solution of baking soda in water exhibits a pH of about 8, as measured by a pH meter. A solution with pH 8 is considered a mild alkali, as the pH is not much higher than a neutral pH of 7.

Most cooks have heard enough about acids, alkalis (also called bases), and the pH scale that they understand this. But what many cooks don't realize is that baking soda is not very stable. As mentioned in the first paragraph, baking soda decomposes with heat to release carbon dioxide gas and water. The other product that is formed by this reaction is called sodium carbonate (see the following chemical equation). Sodium carbonate is a much stronger alkali. A 5 percent solution of sodium carbonate has a pH of more than 11. This means there are about a thousand times more hydroxide ions in the solution of sodium carbonate than there are in the solution of baking soda!

$$2NaHCO_3 \Rightarrow Na_2CO_3 + CO_2 + H_2O$$

But how readily does this decomposition occur? At America's Test Kitchen, I measured the pH of a 5 percent solution of baking soda in water and found it to have a pH of 8. I then gently boiled the solution, keeping the volume constant, while removing small aliquots for testing. After cooling, I measured the pH of the samples with a pH meter. After I had heated the solutions for about 30 minutes, the pH had increased to about 9.7, confirming that much of the sodium bicarbonate had broken down to sodium carbonate, water, and carbon dioxide, which had escaped from the boiling water. It is also possible to get the same reaction in the oven. Heating baking soda at 250°F (121°C) for about 1 hour produces sodium carbonate. As both are white powders, there appears to be no change. But by measuring the pH of solutions of both baking soda and the reaction product, we can quickly confirm the reaction has taken place.

This raises the interesting question of how important this conversion is when cooking and baking with a pinch of added baking soda. Is the effect really due to sodium bicarbonate or to sodium carbonate? Should we even care?

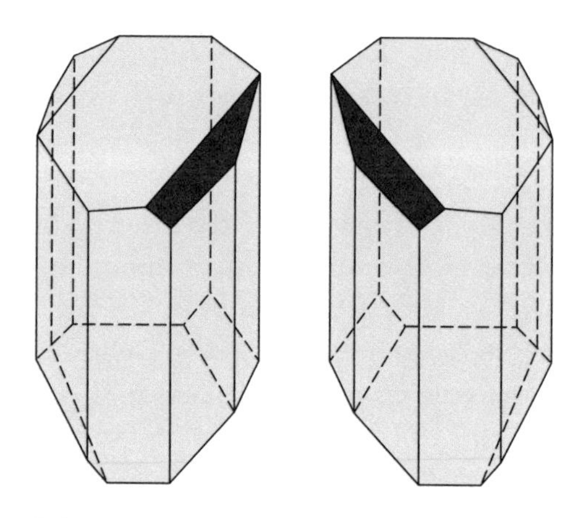

FIGURE 4.3

The left- and right-handed crystals of sodium ammonium tartrate first recognized by Louis Pasteur.

Louis Pasteur (1822–1895) is justifiably renowned for his brilliant research on the role of microorganisms in the fermentation and spoilage of wine and milk, published in 1857. The process of preserving milk and fruit juices by heating them to specific temperatures for specific times to kill the microorganisms was named *pasteurization* in his honor. Far less well known, but of equal scientific importance, is his discovery very early in his career of the three-dimensional shape of naturally occurring organic molecules known as *chirality*. He had a fascination with the shapes of crystals, especially the crystalline salts of tartaric acids obtained from wine. During one of his many attempts to prepare crystals of the sodium ammonium salt of tartaric acid, he noticed that some of the tiny crystals were mirror images of each other, as shown in figure 4.3. The crystal on the right is a mirror image of the crystal on the left. Our two hands bear the same relationship. Our right hand is a mirror image of our left hand, but try as we might, we cannot superimpose either hand on top of the other. That is, placing our right hand directly on top of our left hand, or vice versa, leaves our thumbs on opposite sides.

Pasteur went on to show that solutions of the "right-handed" crystals dissolved in water rotated the plane of polarized light to the right (clockwise), while solutions of the "left-handed" crystals rotated the same plane of polarized light to the left. Thus the mirror-image pairs of crystals were described as being optically active due to their ability to rotate the plane of polarized light either to the left or to the right. When crystals dissolve in water, all the molecules are released into solution. This means the rotation of polarized light that passes through the solution is due to the dissolved molecules and not the crystals. The three-dimensional structure of sodium ammonium tartrate therefore exists as mirror images. One mirror image is shaped like the right hand and the other like the left hand. Pasteur publicized his beautiful research in a lecture at the French Academy of Sciences in May 1848. Today we call molecules that rotate a plane of polarized light in opposite directions *enantiomers*.

Why is this important to the science of food and cooking? Because many molecules that exist in optically active forms produce very different tastes and smells. That's because the protein receptors in our mouth and nose that recognize molecules of taste and smell are themselves optically active and will recognize only one specific enantiomer but not the other. Thus the optical isomer (enantiomer) of monosodium glutamate that rotates polarized light to the right (called *dextrorotatory*) produces the meaty, savory umami taste, while its mirror image, which rotates polarized light to the left (called *levorotatory*), is tasteless. An example related to smell is the optically active, naturally occurring molecule called *carvone*. The enantiomer that rotates polarized light to the right smells like caraway seed, while its mirror image elicits the aroma of spearmint. In this case, the enantiomer that smells like caraway seed presumably activates a different protein receptor from the one that produces the aroma of spearmint. Pasteur's discovery had a profound impact on understanding the chemistry of flavor, as we shall learn later in chapter 5.

A well-known chef who made major contributions to the science of cooking during this period was Joseph Favre (1849–1903). He was a follower of Marie-Antoine Carême and learned early in his career about the importance of applying science to cooking, which enabled the Swiss-born chef to become one of the most famous names in French gastronomy. Orphaned at age fourteen, he was sent to apprentice as a cook with an aristocratic family in Switzerland for three years and then moved to Paris in 1866. Once on his own, he worked in numerous restaurants in France, Germany, Switzerland, and England. In 1877, he launched *La Science culinaire*, a professional journal promoting the application of science to cooking that was successfully published for 7 years. Soon after, in 1879, he founded a professional union for chefs that published another journal, *L'Art culinaire*, thus becoming one of the first chefs to blend both art and science in cooking. Favre also recognized the importance of applying science to create new dishes that were not only appealing and flavorful but also healthy, stating that "culinary science aims to achieve heath through food that sustains virility and the intellectual faculties." He is perhaps best known for writing the four-volume *Dictionnaire universal de cuisine pratique*, published in 1895.

FIGURE 4.4

Mrs. E. E. Kellogg, from the collection of Alfred University, Alfred, New York.

Finally, before the close of the nineteenth century, we come to Mrs. E. E. Kellogg (figure 4.4), one of the first women to write a book on cooking science. Her book, titled *Science in the Kitchen* and published in 1892, focused on cooking methods and recipes for creating healthy food. Emma Eaton graduated from Alfred University (Alfred, New York) in 1872 and shortly thereafter was drawn to work at the Battle Creek Sanitarium, a center for rehabilitating invalid soldiers.

FIGURE 4.5

Cover of *Science in the Kitchen*, by Ella Kellogg, 1892, from the collection of Alfred University, Alfred, New York.

There she met and married the founder, Dr. John Harvey Kellogg, a member of the famous Kellogg cereal family. Emma Kellogg's primary role at the sanitarium was to create and conduct a cooking school, as well as to plan healthy menus for its patients (which numbered 7,000 at its peak in 1906). In 1884, she instituted an experimental kitchen to "develop the principles of healthful cookery," which led her to publish her well-known book. In April 2018, I gave an invited lecture at Alfred University entitled "Science in the Kitchen" and had the good fortune to examine one of the first original copies of Kellogg's book, kept in the university's library (see figure 4.5). Like Favre's journal on culinary science, Kellogg's book established a new direction for cooking science, focusing on understanding the nutritional composition of food and how to select, prepare, and cook food to provide a healthy diet. As you will learn in chapter 6, one of the most important contributions of cooking science is to inform cooks how to prepare food that is not only appealing and flavorful but also healthy. Kellogg was a harbinger of what lay ahead for cooking science and its profound impact on human health.

5 Modern Science Transforms the Art of Cooking (1901–Present)

Flavor Rules

Have you ever noticed how the people who judge cooking competitions on television shows describe the dishes they are rating? For example, "Oh, I like the way you have balanced the acidity with the sweetness of the dish," or "Your dish would have benefited from more contrast between the bitter and salty elements." Do you notice anything? All of the comments relate to the taste of the dish and not the flavor (recall from chapter 1 that taste, smell, and flavor are distinctly different from one another). Sweet, sour, bitter, and salty are all sensations we taste in the mouth; they have no odor. Recall that the image of flavor is created in our brain from what we taste and smell. Extensive research has shown that smell, especially from aromas we inhale into the nose from the back of the mouth, called *retronasal smell*, accounts for perhaps 85 percent of the flavor of food. Yet the judges' comments usually relate to taste and not smell or flavor. Despite perhaps 2 million years of being exposed to the flavor of cooked food, humans have not developed the language to adequately describe smell and flavor. Perhaps this is because we detect only six basic tastes but learn to recognize way more than ten thousand different odors, which combine to create an almost infinite number of flavors in our brain. How can we possibly describe all of them? Supreme Court Justice Potter Stewart described his threshold test for pornography by stating "I know it when I see it." Describing flavor is like that. I know what I like, but I can't really describe it. The people who judge cooking competitions couldn't get away

FIGURE 5.1

A still life of *Peaches in a Glass Bowl*, representing beautiful art and food combined. A pastel by the author (1956), after an original oil painting by A.-F. Bonnardel (1867–1942).

with that; instead, they talk about taste because we have words to describe it. Can you describe the aroma of tarragon other than saying it smells like tarragon? Understanding the science of creating the flavor of cooked food will help us understand flavor even if we can't adequately describe it in words.

There are many reasons why modern humans choose to eat the foods they do, including flavor, nutrition, appearance, texture, safety, and convenience. Many studies conducted around the world have convincingly proved that the flavor of food is the most important determinant of which foods people like and choose to eat and therefore whether their diet is healthy or harmful. In this sense, the flavor of food may have had as much impact on human development as a person's genetics. Creating flavorful, appealing food at home or in restaurants owes as much to the application of cooking science as it does to the quality of the ingredients. Of all the techniques and processes for creating great flavor, none is more important than the Maillard reaction, discovered in 1912 by the French physician and chemist Louis Camille Maillard, a member of the faculty of medicine at the University of Paris. He was interested in diseases of the kidneys and the effects of certain chemical processes that occurred within these two organs. Many amino acids and simple sugars like glucose (from starch or glycogen) and lactose (in milk) are common components of bodily fluids and therefore pass through the kidneys. Maillard decided to see what happens when various amino acids are heated with simple sugars dissolved in water. Perhaps to his surprise, he found that the reaction produces deep brown–colored solutions, which he concluded might be the source of the brown colors formed in cooked foods such as roasted meat and baked bread that also contain amino acids from proteins and sugars from starch or glycogen, the form in which mammals store glucose. Little did he suspect at the time that the reaction subsequently named after him was responsible for some of the most alluring flavors formed when meat is cooked, bread is baked, and coffee and chocolate beans are roasted.

The formation of dark brown colors in heated foods became known as nonenzymatic browning to differentiate it from the formation of brown colors by the action of an enzyme called *polyphenol oxidase* (PPO) in potatoes, apples, and avocados that are sliced and exposed to air. The Maillard reaction takes place between proteins and amino acids and a number of simple sugars (called reducing sugars, a category that does not include sucrose), so it is different from the process of caramelization, which involves the reaction of only sugars and not proteins or amino acids. When cooks refer to browning meat in a hot pan as caramelizing it, they are making a chemical error because the abundant

FIGURE 5.2

John Edward Hodge, USDA chemist.

amount of proteins and amino acids relative to the very small amount of glucose in meat ensures that the Maillard reaction is the source of the color and flavor.

Following Maillard's discovery, other chemists noted the potential contribution of this reaction to the flavor of cooked food. But the true significance of Maillard's reaction did not become apparent until John Edward Hodge (1914–1996), a chemist working for the U.S. Department of Agriculture (USDA), published his extensive observations in 1953. In the most cited paper ever published in the *Journal of Agricultural and Food Chemistry* (a publication of the American Chemical Society), Hodge summarized all of his and prior chemists' research on the role of the Maillard reaction in developing the color and flavor of cooked food. His publication literally turned the world of flavor chemistry upside down. Hodge, an African American born in Kansas City and educated at the University of Kansas, spent 40 years of his life working as a chemist at the USDA laboratories in Peoria, Illinois. Surely he deserves as much recognition for his contributions to flavor chemistry as Louis Camille Maillard.

Thanks to extensive research conducted around the world following Hodge's pioneering publication, we now know that the Maillard reaction can be manipulated in many ways to enhance the flavor of cooked food. For example, although the reaction will take place very slowly even at room temperature, it does not occur rapidly in food until about 302°F (150°C) and higher. That's why foods that are roasted, baked, or seared develop more color and flavor than foods cooked at lower temperatures. Have you ever wondered why most recipes call for baking or roasting in a 350°F oven—and almost never below 300°F (except for barbeque)? It has long been a rule of thumb that the darker the color of roasted meat, the more flavor it will have. Also, the level of moisture in food has a strong influence on the rate of the reaction, with the optimum level being in the middle between very little moisture and very high moisture. Thus the surface of bread dough or a beef roast will not start to brown and develop much flavor until surface moisture has been significantly reduced from the heat of the oven. A steak will start to brown and develop flavor much faster if the surface of the meat has been dried before placing it in the hot pan. The flavor of braised meat is very different from that of roasted meat because the high level of moisture present in the braising liquid discourages the Maillard reaction in favor of other reactions. In addition, a water-based braising liquid will not get any hotter than 212°F (100°C). The pH of the food also plays an important role, as the rate of the reaction increases 500-fold when the pH increases from 5 to 9 (pH 7 being neutral). Cookies that are leavened with alkaline baking soda are much darker than those baked without baking soda. One technique for producing darker, more flavorful skin on roasted chicken or turkey is to rub a very small amount of a weak alkali like baking powder on the skin.

That researchers have identified more than 3,500 flavor compounds formed by the Maillard reaction is proof that the flavor of cooked food is very complex. Only a fraction of these compounds are formed when cooking a specific food like baking a loaf of bread or roasting coffee beans. Many of the compounds are formed in extremely small amounts, yet their threshold for detection (especially through smell) is so low that even minuscule amounts can make a major contribution to flavor. Maillard reaction products can be formed and detected in amounts from a few parts per million to less than one part per trillion. Recall that one part per trillion is the same as 1 second in 32,000 years. That's why some compounds formed in the smallest amounts often make the largest contribution to flavor. In Hodge's day, the identification of Maillard reaction products was a slow, tedious process capable of detecting and identifying only compounds formed in milligram amounts or higher. Sophisticated laboratory instrumentation to separate and identify many of these trace products was not available until the late 1950s, when gas chromatography interfaced with mass spectroscopy (GC-MS) was introduced. The gas chromatograph separates a complex mixture of compounds based on their volatility, which is important for aroma compounds that we detect by smell (recall that aroma is the most important component of flavor). The mass spectrometer determines the mass and structure of each compound. Although mass spectroscopy was developed in 1952, the first computer-controlled MS was not introduced until 1964, and it greatly sped up the process of identifying each compound. GC-MS has been extensively used to identify the trace amounts of flavor compounds in wine and extra virgin olive oil.

Then in the early 1970s, a process called high-performance liquid chromatography (HPLC) was developed and interfaced with MS. HPLC uses a liquid solvent rather than a gas to separate a complex mixture of compounds based on differences in physical properties, thus enabling the separation of nonvolatile compounds that may be important for taste rather than smell. HPLC-MS was more sophisticated in that a method had to be developed to deal with the relatively large volume of water-based or organic solvent introduced into the MS. With computer control, these instruments came to be routinely used for flavor

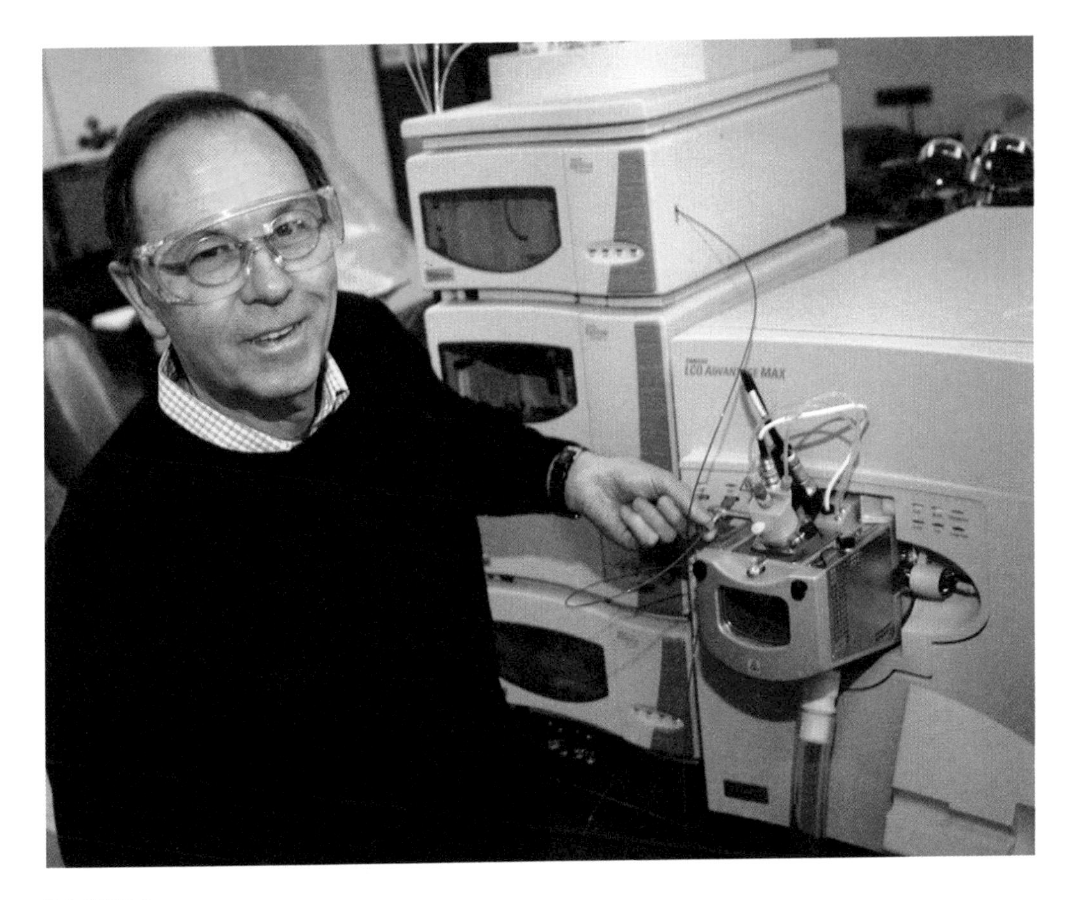

FIGURE 5.3

The author operating a high-performance liquid chromatograph coupled to a mass spectrometer at Framingham State University in 2011.

analysis during the 1980s. Figure 5.3 is a photo of me in front of the HPLC-MS instrument at Framingham State University (the part of the instrument on the right is the MS). It was one of the most challenging instruments I have ever used, even after completing a 1-week training course presented by the manufacturer. Fortunately, during my last 2 years in graduate school, I had a paid teaching assistantship to operate the chemistry department's first mass spectrometer, so I already had some understanding of the basic principles of mass spectroscopy. At Framingham State, we used HPLC-MS to determine the content of a polyphenol called *resveratrol* in wine. With these sophisticated instruments, it is now possible to identify nearly all of the important flavor compounds formed in cooked food.

One variant of the Maillard reaction that has received very little attention involves the reaction of the oxidation products of fats and oils with amino acids and proteins in roasted meat. The oxidation of fats and oils in cooked meat creates much of the delicious flavor through the formation of volatile aldehydes, ketones, alcohols, and a variety of other products. But these oxidation products are also capable of reacting with the abundant amino acids and proteins in meat to form new flavor molecules and colored pigments in a way very similar to the reaction of simple reducing sugars like glucose in the Maillard reaction. One aldehyde in particular, 2,4-decadienal, is formed in relatively large amounts by the oxidation of polyunsaturated fatty acids and has been shown to produce a number of compounds responsible for the flavor of roasted meat via the Maillard pathway. The oxidation of fats and oils requires fairly high temperatures and dry conditions on the surface of the meat, conditions similar to those necessary for the Maillard reaction. Fats and oils that are rich in unsaturated fatty acids, such as those in poultry and grass-fed beef, are more easily oxidized and therefore more reactive with amino acids and proteins than saturated fats are. Although cattle fed cereal grains are higher in total fat, grass-fed cattle contain about five times more unsaturated fat and therefore will produce roasted meat with a flavor profile that is quite different from the meat of cattle raised on cereal grains during the last 120–150 days before slaughter. Rubbing unsaturated cooking oil on the surface of a steak or roast or on the skin of chicken or turkey will enhance both flavor and color.

Vegetables are another food in which flavor can be manipulated for better or worse. There are about thirty-six cruciferous vegetables, now more commonly called *brassicas*, consumed around the world. These include such common vegetables as kale, Brussels sprouts, cauliflower, broccoli, broccoli rabe, and kohlrabi, as well as others less commonly associated with this group, such as arugula, horseradish, and wasabi. They all have one thing in common: their rather sharp, pungent, often bitter flavor when eaten raw. The surprising fact about all these vegetables is that the raw vegetable actually contains *no flavor* at all—at least not until the cells of the vegetable are damaged by cutting, chopping, slicing, or chewing. When I give lectures on the science of taste, smell, and flavor, I often hand out samples of fresh arugula leaves and ask members of the audience to smell them. If the leaves are fresh and undamaged, there is no odor.

Then I ask them to hold the leaf just under their nose and quickly tear the leaf. All of a sudden there is a burst of pungent aroma that surprises them. What happened? When the leaf is torn, cells are ruptured, releasing a compartmentalized enzyme called *myrosinase* that very rapidly reacts with compounds present in the cells called *glucosinolates*. Almost immediately upon contact with the myrosinase, the glucosinolates are converted to volatile pungent compounds called *isothiocyanates*. This occurs within seconds in varying degrees within every cruciferous vegetable. There are dozens of different glucosinolates, so each cruciferous vegetable forms its own unique pungent, bitter flavor when cells are damaged.

Both the glucosinolates and the isothiocyanates are fairly bitter tasting, so all of the vegetables in this group have this pronounced taste in common. The more damage that is done to the cells, the more flavor that is developed, so finely chopping the vegetable instead of slicing it creates more flavor. Allowing the chopped vegetable to rest for 5 or 10 minutes before cooking also increases the amount of flavor. But many people, especially young children, dislike the bitter taste of cruciferous vegetables like broccoli. Blanching the vegetable for about 30 seconds in simmering water deactivates most of the myrosinase (which becomes inactive above about 140°F, or 60°C). Cooking the vegetable after blanching significantly reduces the bitter taste. Try this with your children to see if it increases their liking of cruciferous vegetables. Finally, the isothiocyanates are molecules that contain a sulfur atom. As chopped or sliced cruciferous vegetables are cooked, the pungent isothiocyanate molecules are slowly converted to more mellow, nutty-flavored sulfur compounds called *disulfides* and *trisulfides*. At America's Test Kitchen, the test cooks cooked cauliflower in boiling water for different periods of time and asked panelists to taste the cooked vegetable at different intervals. They found that, after cooking for only 10–20 minutes, the cauliflower was still sulfurous tasting due to the formation of volatile hydrogen sulfide. After 30–40 minutes of cooking, the cauliflower had the best mellow, nutty flavor, while after 50–60 minutes, the flavor was exceptionally bland. Soup made by cooking the cauliflower for 30–40 minutes clearly had the best flavor. Because all cruciferous vegetables owe their flavor to sulfur-containing compounds, the sulfur content of the soil in which they are grown is very important, with higher sulfur content providing more flavor. This is an example of what the French call *terroir*, meaning the factors in the natural environment where wine grapes are grown that affect the flavor of the wine. But terroir can be just as important in terms of the flavor of plant foods like the cruciferous vegetables.

The French are given credit for being the first to recognize the importance of the taste of place, or *terroir*. They have long identified the place of growing food with the qualities of that food. For this reason, French wines are named for the area where the grapes are grown rather than for the variety of grape, as is done in the United States. It is now widely recognized that terroir extends beyond wine; the quality and flavor characteristics of olive oil, cheese, and honey are related to where the olive trees, cows, and honeybees live and grow. Certain wines and other agricultural products are labeled *Appellation d'Origine Contrôlée* in parts of France and *Denominazione di Origine Controllata* in Italy to indicate they are produced in specific regions of these countries. In the United States, we have Vidalia onions, which may be grown only in the southeastern counties of Georgia, according to the Vidalia Onion Act of 1986. For an excellent review of the role of terroir in the quality of food, read *The Taste of Place* by Amy B. Trubeck.

Cooking with local ingredients is becoming a huge trend in the United States. Most chefs believe that local ingredients not only are fresher but also taste better. But is there any science to support this? Yes! There are now many examples in the scientific literature of the impact of growing conditions and the environment on the flavor and texture of food. Perhaps best known among these foods are the alliums, which include onions, garlic, leeks, scallions, shallots, chives, and even ramps. The alliums develop their characteristic flavor and pungency only when the enzyme *alliinase*, released when their cells are damaged, comes in contact with sulfur-containing compounds called S-alkenyl cysteine sulfoxides (ACSO). That's why a whole head of garlic or an onion bulb has no odor. The more that onion and garlic are chopped, the more intense their flavor and pungency are. The ACSO compounds are derived from the natural sulfur-containing amino acid called *cysteine*. Not surprisingly, research studies have shown that the intensity of the flavor and pungency of alliums is related to the sulfur content of the soil, which is referred to as the *sulfur fertility* of the soil. Sulfur exists in the soil in the form of sulfate salts, which are absorbed by the plants and converted into cysteine and a few other sulfur-containing amino acids. Higher levels of sulfate in the soil are correlated with more flavor and pungency. Milder, sweeter onions like the Vidalia onion are grown in soils with low levels of sulfate.

Another class of vegetables, the cruciferous vegetables, produce their characteristic flavor and pungency through chemistry that is similar to that of the alliums. But in this case, an enzyme called *myrosinase* is liberated when the cells are damaged. Myrosinase converts a different group of sulfur-containing compounds called *glucosinolates* into the compounds called *isothiocyanates* that are responsible for the flavor and pungency of cruciferous vegetables. There are about thirty-six different cruciferous vegetables commonly consumed around the world. In the United States, the most common are cabbage, kale, Brussels sprouts, broccoli, broccoli rabe, cauliflower, mustard greens, collard greens, turnips, bok choy, Swiss chard, radishes, and arugula. All these are tasteless until chopped, sliced, or chewed. More cell damage creates more flavor. Blanching cruciferous vegetables inactivates myrosinase, resulting in milder, less bitter flavor. Research has shown that, as with alliums, the sulfur fertility of the soil correlates with the flavor and pungency of cruciferous vegetables.

What really got me thinking more about the validity of the taste of place was a blind taste test done with cooked dry cannellini beans at America's Test Kitchen in 2013. Beans from five different sources were tasted plain, in a dip, and in a soup. All of the beans were tasted blind six times by a panel of twenty-one tasters each time. The dry beans from each source were also sent to a laboratory to determine their levels of calcium. Calcium in beans is mostly associated with the pectin that holds the cells together and strengthens the cell walls. It was anticipated that more calcium would result in less "blow out" (bursting of the outer shell) of the beans and a better interior texture. The results were astounding! As shown in the following table, the ranking of the beans by taste and texture correlated perfectly with the calcium content, with the beans containing the most calcium ranking the highest. There were even two brands of beans that contained essentially the same amount of calcium that tied in the taste ranking! As you might have guessed by now, research studies have shown that the level of calcium in beans is related to both the genotype and the calcium content of the soil. So, again, where the beans are grown matters!

Cooked Dried Cannellini Beans

Rank	Brand	Calcium (mg/100g)
1	A	362
2	B	204
Tie	C	176
Tie	D	175
5	E	168

Outside of the world of plant foods, I should mention the interesting exploration of *microbial terroir* by David Chang, chef/owner of the Momofuku restaurant group. Chang is experimenting with local microorganisms to impart different regional flavors to fermented foods. Benjamin Wolfe, who collaborated with Chang while working at Harvard University and who is now at Tufts University, has evaluated the different bacteria, yeast, and mold that colonize regionally made salamis, and his results are fascinating. According to Wolfe, salamis colonized with different microorganisms have distinctly different flavors. Notice in figure 1 the different local yeasts and molds growing on the surface of the salamis. Microbial terroir at work!

Finally, the question of the taste of place arises as to other foods. It is easy to see why the flavor of alliums and cruciferous vegetables is closely linked to where they are grown. They owe their flavor to specific enzymatic reactions that occur with a small group of sulfur-containing

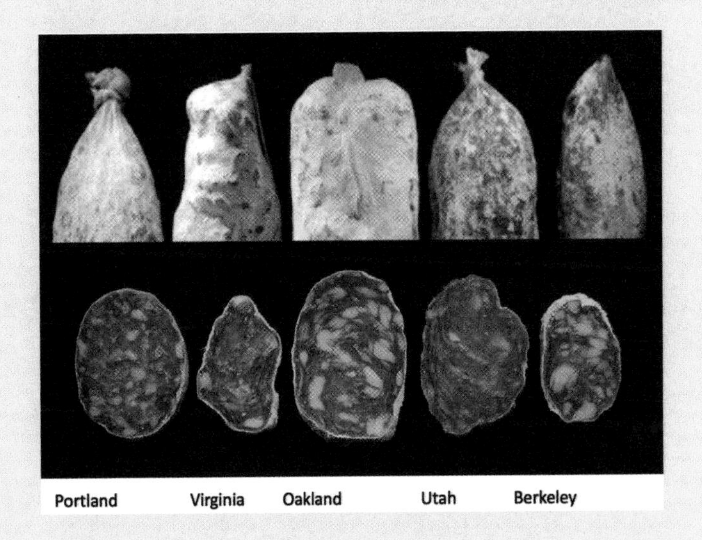

FIGURE 1

A comparison of the insides and outsides of different samples of salami produced in Portland, Oregon; Virginia; Oakland, California; Utah; and Berkeley, California, illustrating how local microorganisms produce different forms of salami. Photographs provided by Benjamin Wolfe, Tufts University.

compounds derived from the amino acid cysteine. So the sulfur content of the soil is a very important factor in the intensity of the flavor of these plant foods. But many other vegetables have more-complex sources of flavor. Potatoes are an example. Potatoes that are boiled, baked, or steamed owe their flavor to many different reactions that occur during cooking, including the reactions of fats and sugars and amino acids, the formation of sulfur-containing compounds, and the Maillard-Hodge reaction, which forms potent aroma compounds called *pyrazines*. Research has shown that the flavor of cooked potatoes is determined by both genotype and environmental conditions. But because of the complexity of the many flavor-forming reactions, there is no single environmental factor, such as the sulfur content of the soil, that can be correlated with cooked potato flavor. Not all foods can be linked to a taste of place.

REFERENCES

Boyhan, G. E., and R. L. Torrance. "Vidalia Onions—Sweet Onion Production in Southeastern Georgia." *HortTechnology* 12, no. 2 (2002): 196–202.

Falk, K. L., J. G. Tokuhisa, and J. Gershenzon. "The Effect of Sulfur Nutrition on Plant Glucosinolate Content: Physiology and Molecular Mechanisms." *Plant Biology* 9 (2007): 573–581.

Felder, D., D. Burns, and D. Chang. "Defining Microbial Terroir: The Use of Native Fungi for the Study of Traditional Fermentative Processes." *International Journal of Gastronomy and Food Science* 1 (2012): 64–69.

Quintana, J. M., H. C. Harrison, J. Nienhuls, J. P. Palta, and K. Kmiecik. "Differences in Pod Calcium Concentration for Eight Snap Bean and Dry Bean Cultivars." *HortScience* 34, no. 5 (1999): 932–934.

Randle, W. M., D. E. Kopsell, and D. A. Kopsell. "Sequentially Reducing Sulfate Fertility During Onion Growth and Development Affects Bulb Flavor at Harvest." *HortScience* 37, no. 1 (2002): 118–121.

Trubeck, A. B. *The Taste of Place—A Cultural Journey Into Terroir.* Berkeley: University of California Press, 2008.

This brings us to the flavorful alliums, which include garlic, onions, leeks, shallots, chives, and scallions (often referred to as the onion family). Like the cruciferous vegetables, the fresh, intact alliums contain *no flavor* until their cells are damaged. To prove the point, smell a fresh whole head of garlic or an undamaged onion or shallot, and you will smell nothing. When the cells of alliums are damaged, they release a compartmentalized enzyme called *allinase*, which converts odorless sulfur-containing amino acids called *cysteine sulfoxides* to volatile pungent flavor compounds called *thiosulfinates*. More cell damage creates more flavor, so mincing garlic produces more flavor than slicing or chopping it. To enhance the flavor of alliums, they should be placed in room-temperature oil and heated rather than being tossed into hot oil. This gives the enzyme more time to form the flavor molecules before the heat deactivates it. Again, like with cruciferous vegetables, sulfur-containing compounds play a very important role in the flavor of alliums, so the sulfur content of the soil where the alliums are grown is very important. Mild, sweet-tasting Vidalia onions are grown in regions of Georgia with sulfur-poor soils—yet another example of how terroir affects the flavor of many foods. Cooking also mellows the sharp, pungent flavor of the alliums by converting the thiosulfinates to disulfides and trisulfides. But be careful not to overcook garlic; it produces bitter-tasting compounds when it begins to turn brown in hot oil.

Although all the alliums have very similar flavor chemistries, onions, shallots, leeks, and chives differ from garlic in one very important aspect: they contain an additional enzyme that is not present in garlic. This enzyme, called *lachrymator factor synthase* (LF synthase), converts the pungent thiosulfinates to propanethial S-oxide (PSO), a volatile compound known as a lachrymator because it causes the eyes to tear. The more cell damage that is done, the more PSO that is formed, so mincing onion creates more tears than slicing onions. In fact, slicing an onion from pole to pole damages fewer cells and produces less tearing than slicing an onion through the middle. Finally, chopping or mincing onions might be bad when it comes to producing tears, but it has a definite advantage when cooking members of this family for their flavor. Relatively recent research has shown that cooking finely chopped onion in water for 1 or 2 hours, or longer, produces a savory, beefy-tasting compound called 3-mercapto-2-methylpentan-1-ol (MMP). The name of this simple sulfur-containing compound may sound complicated, but it describes the chemical structure of the compound, as do the names of most organic compounds (see the chemical structure in figure 5.4). Although MMP is formed in extremely small amounts (about 50 micrograms per kilogram of yellow onions; a microgram is one-millionth of a gram), it has been shown to have the greatest impact of all flavor compounds formed in vegetable and beef stocks and gravies prepared with

(CONTINUED ON PAGE 103)

RECIPE 5.1 Julia Child's Brown-Braised Onions

INGREDIENTS:

1½ Tbsp. clarified butter

1½ Tbsp. extra virgin olive oil

24 small white pearl onions, 1–1½ inches in diameter, peeled

½ cup beef stock or bouillon

½ cup dry light red wine, such as pinot noir or Beaujolais

4 parsley sprigs

½ bay leaf

¼ tsp. dried thyme

½ cup light cream, optional

NOTE:

The recipe can be scaled up two or three times using one or two 12-inch skillets. Christine has made as many as sixty to sixty-five onions at a time for large Thanksgiving dinners.

***Yield*: 6 servings (see note)**

Julia Child's recipe for brown-braised white pearl onions has been a favorite at our Thanksgiving table for decades, during which time it has been prepared with loving care with relatively minor variations by my wife, Christine. The intensity of the onion flavor, augmented by the subtle aroma of selected herbs and wine, brings back memories and sensations that are hard to surpass on a beautiful crisp fall day in New England. Although Child originally developed the recipe to accompany coq au vin or boeuf bourguignon, we find the onions pair extremely well with oven-roasted turkey stuffed with chestnuts and with all the other dishes at our traditional Thanksgiving table, including deep orange butternut squash, creamy mashed potatoes, bright green steamed beans, homemade cranberry-orange relish, and especially Christine's incredibly rich brown gravy made from slowly simmered turkey stock.

There are so many complex facets to the intense flavor of these melt-in-your-mouth onions. The first stage of flavor development arises when the onions are browned in a mixture of extra virgin olive oil and clarified butter, resulting in both the caramelization of the sugars and the Maillard reaction of the sugars with proteins in the butter and onion. The butter should be clarified so it reaches the high temperature required for both reactions without being delayed by evaporating the water in the butter. Onions are rich in a nonvolatile, non-odorous, organic sulfur-containing amino acid known as 1-propenyl-L-cysteine sulfoxide (1-PRENSCO), which they produce from inorganic sulfate absorbed from the soil.

The second stage of slowly cooking the browned onions in a watery liquid such as wine, stock, or even plain water converts the nonvolatile sulfur compound to volatile potent flavor compounds such as savory propenyl propyl disulfide and meaty-tasting 3-mercapto-2-methylpentane-1-ol (MMP), which humans can smell at levels as low as 0.1 to 0.01 parts per billion. In addition, onions contain relatively high levels of both glutamate and nucleotides that interact synergistically to produce a savory umami taste. In fact, onions contain about 70 percent as much glutamate and nucleotides as shiitake mushrooms, which are a well-known source of great umami taste. The formation of these intensely savory compounds in the slowly cooking onions, combined with the aromas and tastes released from the wine and herbs, produces an intense, complex flavor that is hard to surpass.

DIRECTIONS:

Heat the butter and olive oil in a 10-inch stainless steel skillet on medium heat until shimmering (but not smoking). Add the onions, and cook over medium heat for about 15 minutes, rolling the onions in the butter–olive oil mixture from time to time until they are uniformly brown. Add the beef stock or bouillon, red wine, parsley, bay leaf, thyme, and a little salt to the pan, and mix. Loosely cover the pan and simmer very slowly for about 1½ hours, until the liquid has reduced to a thick syrup. If necessary, remove the cover to allow more rapid evaporation of the liquid. Check frequently, and stir the onions to make sure they do not burn as the liquid is reduced. If desired, add about ½ cup of light cream to produce a creamier sauce. The onions can be refrigerated and reheated several days later.

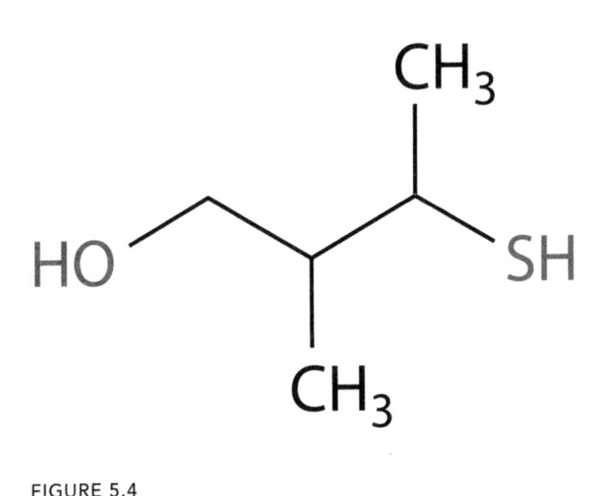

FIGURE 5.4

The chemical structure of 3-mercapto-2-methylpentan-1-ol (MMP).

members of the onion family (MMP has a recognition threshold of only 0.0016 micrograms per liter in water). Speaking from personal experience, the slow formation of this water-soluble compound has a tremendous impact on the savory, beefy flavor of stock and gravy. Thus, for more intense flavor, stocks should be prepared with finely chopped onions and *not* whole onions. This explains why cooking stocks for long periods (hours) with finely chopped onion develops more flavor and why cooking onions in water develops more savory flavor than sautéing them in oil, which forms more caramelized flavor compounds. MMP is formed at levels five to seven times higher in leeks and chives than in yellow onion, which is why these members of the onion family form more flavorful stocks than simple yellow onions, the frequent choice of many cooks.

Modern Methods of Cooking

Food science received very little academic attention before the early 1900s, resulting in a dearth of basic research to support the science of cooking. What little research there was had been conducted by departments of home economics, also known as domestic science. Early research efforts focused on plant and animal sciences and supported agriculture rather than food or cooking. By 1905, ten dairy schools had been established in the United States and Canada, including those at the land-grant colleges of Cornell, Wisconsin, Iowa State, Purdue, and Penn State, but there were no departments of food science. The American Chemical Society created its Division of Agricultural and Food Chemistry in 1908, and the University of Massachusetts at Amherst established the first department of food science in 1918. In 1936, the journal *Food Research* was launched, and shortly thereafter it was purchased by the Institute of Food Technologists (founded in 1939) and renamed the *Journal of Food Science*. Since these early efforts, food science has become an established academic discipline, drawing on research in chemistry, biology, physics, engineering, microbiology, and nutrition. As of 2017, there were forty-five departments or schools of food science in the United States, with many combined with departments of nutrition. Research conducted at these institutions has provided the bulk of our current knowledge of the science of cooking for the past 100 years, yet throughout all these years, departments of food science have paid relatively little attention to cooking science as an independent field of study.

The three basic methods of cooking—beginning with dry forms of heat such as roasting and grilling, followed by wet methods such as boiling and braising, and then by frying or sautéing in fat or oil—have been in use for at least 5,000 years. Even sous vide cooking at a constant low temperature traces its roots back to 1799. Yet in all this time, the only other truly new method of cooking to appear has involved the use of microwave ovens, developed by Percy Spencer at the Raytheon Corporation in 1945. The use of microwaves for heating food was based on radar technology that Raytheon developed during World War II. The first commercial "Radarange," introduced in 1947, was the size of today's refrigerator and intended for commercial use. The countertop microwave oven, first sold by the Amana Corporation in 1967, proved to be so popular that by 1997 more than 90 percent of American households owned a microwave oven!

Microwave ovens are great for quickly heating or reheating food, but they cannot brown meat or make crispy fried chicken. Basically, microwaves cause the excitation and vigorous movement of water molecules in food, thus cooking food at no higher than 212°F (100°C), a temperature that is too low to rapidly brown or crisp food. Their other limitation is the uneven distribution of microwaves in the enclosed oven, causing cold spots in foods like a whole russet potato or a sizable piece of meat. In the brief time needed to "microwave" foods, the electromagnetic waves penetrate only about the outer inch of the food (depending on the food) before they are all absorbed by water molecules; the interior of the food is cooked as conduction slowly transfers the heat from the hot surface into the food. Despite these limitations, the convenience of rapidly heating foods has made microwave ovens a huge success.

Sous vide cooking is often presented as a modern invention, first used in 1974 by French chef Georges Pralus at the Troisgros Restaurant in Roanne, France (*sous vide* is a French term meaning "under vacuum"). But we learned in chapter 3 that Benjamin Thompson actually demonstrated precise low-temperature slow cooking of a mutton roast in 1799 using a device for dehydrating potatoes, which he also invented. Thompson did not seal food under vacuum in plastic pouches, as is done today, so technically his method does not meet the definition of sous vide cooking, but it clearly demonstrated that cooking meat at a low, controlled temperature for a very long time rendered even tough cuts of meat tender, juicy, and flavorful. The sous vide process, in which food is vacuum sealed in a plastic pouch and cooked slowly at a precise low temperature, retains the moisture and nutrients in meat or fish while also slowly breaking down tough connective tissue to gelatin. Sous vide cooking is now widely used in many high-end restaurants. It is especially suited for slowly par-cooking all sorts of foods, rapidly freezing the food in vacuum-sealed plastic pouches for many hours or overnight, and then quickly finishing the cooking process when the customer's order is received in the kitchen. It is reported that some international food service organizations cook large batches of foods by sous vide in a central location, flash freeze them, and then ship them all over the world to be finished to order in local restaurants and cafeterias.

It would be an oversight not to include in the realm of modern methods of cooking what has become known as *molecular gastronomy*—and more recently called *molecular cuisine*—which has been defined as "the scientific discipline that deals with the development, creation, and properties of foods normally prepared in the kitchen." I prefer the term *molecular cuisine* and define it as the application of science for the purpose of creating outstanding food with unique flavor, texture, and appearance. Clearly, it is not, as some have suggested, merely a style of cooking, like fusion cuisine. The creation of this science-based approach to cooking traces its roots to a conference on this subject held in Erice, Italy, in 1992. Although many well-known chefs have embraced molecular gastronomy, some have moved away from this new approach to cooking for fear of appearing to use the method to create gimmicks rather than to prepare great food. Many serious diners and chefs find the concept of molecular gastronomy to be confusing and even pretentious, perhaps because *gastronomy* is defined in my dictionary as both "the art or science of good eating" and "cooking, as of a particular region." Which is it, cooking or eating? Art or science? For this reason, I prefer the less ambiguous word *cuisine*. Add the word *molecular* in front of *gastronomy* or *cuisine*, and many are left wondering if it represents the missing link between art and science or a clever marketing ploy. But those who practice molecular gastronomy or molecular cuisine for the purpose of creating great food are serious advocates for applying science to the art of cooking. I must state emphatically that the field of cooking science that is the subject of this book is far broader and more comprehensive than molecular cuisine.

Many of the world's top trend-setting chefs—such as Alice Waters, Ferran Adrià, Heston Blumenthal, René Redzepi, Wylie Dufresne, Thomas Keller, and Grant Achatz—have utilized the concepts of molecular cuisine, confirming this modern approach to cooking should not be taken lightly. Yet they are adamant that molecular cuisine does not define or limit their approach to cooking, which is far broader and more imaginative and creative. It is only one tool in their vast portfolio of tools. These great chefs have elevated the texture and appearance of food to the same level as flavor, creating truly novel forms of foods that are a feast for the eyes, nose, and mouth. Edible gels, foams, and emulsions have been used in cooking for hundreds of years, but the science of how they work has been understood for only about the last 100 years. In my earlier career, I worked for the world's largest producer of the polysaccharide carrageenan, isolated from different varieties of red seaweed. Carrageenan is one of the gelling agents used in molecular cuisine. FMC Corporation started manufacturing carrageenan back in the 1940s and understood a lot about the science of gels. We were also using carrageenan and its cousin sodium alginate in the early 1980s to make spherical gel beads, similar to the products of spherification, a creative technique popularized more recently by Ferran Adrià. But molecular cuisine is not only about creating new forms of food using gels, foams, and emulsions; special equipment like rotary evaporators; and novel substances like liquid nitrogen. Innovative chefs have combined their knowledge of science with the art of cooking to create truly inspirational new forms of food, often based on the highest-quality seasonal local ingredients. These chefs are like emulsifiers, blending art and science into gastronomic delights. But for me, a simple dessert of very fresh strawberries, served with a dollop of sour cream sprinkled with brown sugar, is just as good, if not better, than the most imaginative dessert created using molecular gastronomy. This simple fresh dessert was first served to me many years ago by my late dearest friend, Bill Benz, and his wife, Ruth.

Today modern, globe-trotting restaurant chefs have moved well beyond molecular gastronomy by exploring hybrid cuisines based on their diverse heritages and cultural backgrounds, producing a fusion of ingredients and bold flavors from around the world. Great food and cooking are no longer dominated by a single cuisine as they were during the heyday of French cooking. Like so many things today, cooking has become globalized, and science has helped to guide this change. Take, for example, the innovative cooking of David Chang, chef/owner of many successful restaurants, including Momofuku in New York City. Chang has pioneered the use of fermentation to create unique new foods and flavors by collaborating with microbiologists and even publishing his research in academic journals. His team has used native (local) forms of environmental microbes such as fungi and bacteria to increase the amino acid content and umami taste of fermented meats and grains, naming this novel approach *microbial terroir*. After researching the preparation of *katsuobushi* (dried, fermented, and smoked skipjack tuna, known as bonito flakes), the team developed *butabushi*, a new form of fermented steamed pork, as well as new forms of *koji* (fermented rice) and *miso* (fermented soybeans). This work led to the creation of three new sauces—Hozon, Bonji, and Ssäm—derived from the fermentation of grains, nuts, seeds, and legumes other than soybeans.

SOFT MATTER—THE SCIENCE OF GELS

Physics involves the study of matter, and one form of matter that is important to food and cooking is *soft matter*, which includes foams and gels but also extends to liquids, colloidal suspensions, emulsions, and polymers. Among the common foams and gels are ice cream, meringues, mousses, puddings, pie fillings, and flans. Emulsions are well known in the form of mayonnaise, dressings, and sauces. The advent of molecular cuisine led to the creation of novel forms of foams and gels based on an understanding of the science of soft matter. Even though the study of soft matter is relatively new, gels have been in existence for a very long time. Gelling and thickening agents derived from connective tissue have been used in home cooking for more than 2,000 years. The name *gelatin*—based on the Latin word *gelatus*, meaning "firm or frozen"—was in common use as early as 1700. The commercial production of purified gelatin began in the United States in 1850 and in Europe about twenty-five to thirty years later.

Gelatin is derived from the principal protein in connective tissue, called *collagen*, an aggregate of proteins that exist in the form of strong fibers. Treating sources rich in collagen, especially animal hides, bones, and pigskin, with either acid or alkali breaks down the collagen fibers into a single protein called *tropocollagen*, which consists of three very similar protein chains wound into a triple helix. Further processing with heat unwinds the triple helix into single strands of the protein known as *gelatin*. When placed in cold water, gelatin swells, absorbing as much as five to ten times its weight of water; it then dissolves at higher temperatures and upon cooling to 81°F–93°F (27°C–34°C) forms a relatively soft but elastic gel that melts in the mouth. The strength of a gel, known as the *bloom value*, is determined by measuring the weight in grams required to move a piston into the gel (made with about 7 percent gelatin) to a depth of 4 millimeters. Gelatin has many applications in food, including confectionary and jelly desserts; dairy products such as yogurt; meat products, where it replaces fat; and sauces and dressings. Perhaps gelatin's most useful application is in cooking sauces; when added, it produces a luscious, unctuous mouthfeel similar to that obtained by reducing veal stock to a demi-glace. The collagen in veal (especially the bones) breaks down to gelatin much more readily and at much lower temperatures than that in tough cuts of beef or pork, which must be cooked for hours to obtain the same amount of gelatin.

Gelatin is a very large protein molecule that forms regions of helices when solutions cool. The helical regions can be formed within a single gelatin molecule or by several molecules intertwining to form a very large network held together by double or triple helices that trap water, forming a soft gel structure. Substances that form gels with water are known as *hydrocolloids*. The general mechanism of large molecules entangling with each other to trap water within the network of molecules (but not necessarily by forming helices) is found in other gelling agents, such carrageenan, pectins, alginates, xanthan gum, gellan gum, and starch (discussed in the "Microscopic World of Starch in Food" in chapter 1). In all of these cases, the molecules are very large polysaccharides (polymers of small sugar molecules) rather than proteins.

Various forms of red seaweed have been used in food in China going as far back as 2,600 years ago. Carrageenan has been produced commercially since the 1930s and is most commonly obtained by treating different varieties of red seaweed with alkali and then precipitating the product with alcohol or potassium chloride. There are three forms of carrageenan—*kappa*, *iota*, and *lambda*—that are obtained from different varieties of red seaweed. Kappa carrageenan forms strong brittle gels in the presence of potassium and calcium ions, iota carrageenan forms weak elastic gels with calcium ions, and lambda carrageenan acts to thicken water but does not form a gel in the presence of ions. The ability to form gels in the presence of calcium ions makes very low concentrations of kappa and iota carrageenans useful for producing milk-based gels such as puddings, flans, and yogurts. Carrageenan also interacts synergistically with nongelling food gums, such as guar gum and locust bean gum, to produce gels with very unique properties. With its variety of different properties from gelling to thickening, carrageenan has found numerous uses in meat, frozen desserts, water gels, many dairy products (including ice cream, cheese, and chocolate milk), toothpaste, salad dressings, beverages, coffee creamers, and pet food.

Pectin, another polysaccharide hydrocolloid, was first isolated in relatively pure form in 1825 and has been in use for a very long time, starting with the home preparation of fruit preserves. Pectin occurs naturally in most plant materials, where it forms part of the cell walls and holds the plant cells together, and is most abundant in citrus fruits and apples. Some fruits like strawberries contain very low levels of pectin, which explains why strawberries must have pectin added to them in order to form strawberry jam and jelly. There are two forms of commercial pectin, low-methoxyl pectin and high-methoxyl pectin. The latter form will produce strong gels at relatively low pH (pH 3.2–3.4 is optimum) but only in the presence of high concentrations of sugar, which is why fruit preserves contain so much added sugar. Low-methoxyl pectin forms gels in the presence of calcium ions, which act as bridges to hold molecules of pectin together into a network, but it also requires relatively low pH. The acid is required to alter the electrical charges on the pectin molecules, helping them to bond together. Low-methoxyl pectin is used not only in jams, jellies, and preserves but also in a variety of milk-based products, including yogurt and cultured milks.

Gels made with sodium alginate have been used commercially since about 1940 and are commonly encountered in processed food. Sodium alginate is a chemical relative of carrageenan but is found in brown seaweeds rather than the red varieties. It forms gels in the presence of added calcium ions in the same manner as carrageenan: the calcium ions act as bridges that link molecules of alginate together to trap water within the infinite network of molecules. Unlike carrageenan gels, alginate gels are heat stable, thus finding numerous applications in restructured foods such as onion rings, pimento fillings for olives, fish patties, and pet food, as well as in bake-stable filling creams, salad dressings, mayonnaise, ketchup, and instant puddings.

A more recent addition to the list of food-approved hydrocolloids is xanthan gum, discovered by the U.S. Department of Agriculture. Commercial production of xanthan gum started in the United States in the 1960s, and its use in food was approved by the Food and Drug Administration (FDA) in 1969. Xanthan gum differs from the complex polysaccharides discussed earlier in that it is produced by fermentation of a bacterium, *Xanthomonas campestris*, rather than by isolation from plant material. Strictly speaking, xanthan gum is not a gelling agent. Rather, it forms very viscous solutions at very low concentrations and interacts with other gums, such as guar gum, locust bean gum, and sodium alginate, to produce solutions with unique thickening properties, such as reaction and stability to high-speed mixing (called *shear*) and resistance to acids, bases, proteins, sugars, and salts. As a result of these properties, xanthan gum has found applications in salad dressings, sauces, gravies, canned soups, ice cream (to reduce ice crystal formation), syrups, and fillings.

The most recent gel-forming food-approved polysaccharide is gellan gum, produced by aerobic fermentation of the bacterium *Pseudomonas elodea* and patented in the 1980s. Gellan gum forms very viscous solutions in cold water, and when heated to 176°F–194°F (80°C–90°C) and then cooled to room temperature, it forms gels at concentrations as low as 0.05 percent. Gel strength is enhanced by the addition of various gel-forming salts containing calcium and sodium ions. By regulating concentrations of both gellan gum and different salts, gels can be produced with properties ranging from soft and elastic to hard and brittle. Applications range from water-based jellies to fabricated foods, icings, milk-based desserts, and pie fillings.

During my last 10 years working in the food ingredients industry, I was involved with a very interesting food hydrocolloid called *konjac flour*, a complex polysaccharide isolated from tubers of *Amorphophallus konjac*, a plant native to China, Japan, and several other Asian countries. Konjac flour has been known and used in China and Japan for many years not only for its unique gelling properties in food but also for its ability to lower cholesterol levels in the blood. When a viscous 1 percent solution of konjac flour in water is treated with a mild alkali, such as potassium carbonate or even ground oyster shells (primarily calcium carbonate), and heated to about 176°F (80°C) for a few minutes, it forms a firm, very elastic, rubbery gel upon cooling to room temperature. The gel is amazingly heat stable, and if placed on the surface of a hot plate heated to 350°F (177°C), it will remain for several hours without melting or drying out. But even more amazing is that, when the gel is cooled to near freezing (about 39°F, or 4°C), it turns back into a viscous liquid! Konjac flour is approved by the FDA for use in providing heat stability in veggie burgers, meat patties, surimi (imitation fish), and noodles.

REFERENCE

Imeson, A., ed. *Thickening and Gelling Agents for Food*. Glasgow: Blackie Academic and Professional Press, 1992.

STAR POWER

Star anise is an important culinary spice used in Asian cooking to flavor slow-cooked braised and stewed dishes. It is often used both by itself and as a component of the well-known Chinese five-spice blend that typically consists of finely ground anise pepper, star anise, cassia, cloves, and fennel seed. The five-spice blend is used in marinades and barbeque sauces, on roasts, and in stews and soups such as the Vietnamese beef noodle soup known as *pho*. Chinese cooks use star anise to season braised and stewed pork and stewed chicken dishes that are cooked for several hours in a blend of four parts water and one part soy sauce seasoned with star anise. This is one of their favorite ways to cook delicious braised fatty pork belly. Although star anise was brought to Europe from China near the end of the sixteenth century, it has found relatively little use in Western cooking even today.

Star anise is the fruit of an evergreen tree (*Illicium verum* Hook) that is native to southwestern China and northern Vietnam and now cultivated in these countries, as well as in Laos, India, the Philippines, Japan, and Korea. As its name implies, the fruit is an eight-pronged star-shaped pod. The pods are harvested when still green, but they turn to a reddish-brown color after drying in the sun and develop a licorice-like, sweet, but pungent flavor that intensifies during prolonged cooking in a stew or broth. The dried pods of star anise are about 2.5–3.5 percent essential oils by weight. A volatile compound called *trans-anethole* makes up about 85–90 percent of the oils, another volatile compound called *methyl chavicol* makes up about 2 percent, and numerous other compounds are present in smaller amounts. Pure trans-anethole is a crystalline semisolid that melts at about room temperature (70°F, or 21.4°C) and is also found in star anise, anise, and fennel oils, providing the characteristic licorice-like flavors of all three oils. The oils are soluble in fat and alcohol but very poorly soluble in water. Interestingly, trans-anethole is a sweet-tasting compound that is about thirteen times sweeter than sugar.

After prolonged heating in water, trans-anethole is slowly converted to an oily liquid named 4-methoxybenzaldehyde (also known as p-anisaldehyde) that possesses a fairly pleasant vanilla-like aroma. When it comes to flavor development in slow-cooked braised and stewed meats and poultry, this compound plays one of the most significant roles and is the reason why the stews and braises must be cooked for several hours to develop the desired flavor. This brings up an important point in the development of cooked meat flavor. What little flavor raw meat has comes mostly from the taste of the residual blood in the meat. The characteristic flavor of cooked meat arises from chemical reactions that occur during cooking. One of the most important sources of flavor is the Maillard reaction between sugars, such as glucose and fructose, and various amino acids formed by the breakdown of proteins. But this reaction occurs much more readily when meats with very limited water are roasted at temperatures above 300°F (about 150°C) and can be easily seen, for example, on the hot, dry surface of a roast. Boiling meat produces very different flavors because the Maillard reaction is more subdued in high concentrations of simmering water, which causes very different chemical reactions than dry roasting.

The process of slowly simmering meat is where star anise makes such a large contribution to flavor. Research recently conducted at Rutgers University's Department of Food Science has shown that trans-anethole slowly breaks down to 4-methoxybenzaldehyde in simmering water, which in turn reacts with the sulfur-containing amino acids L-cysteine and cystine to produce the characteristic flavors of Chinese stewed pork dishes. The primary product of the reaction with L-cysteine is a sulfur-containing compound called 4-methoxybenzothialdehyde, which is responsible for much of the stewed meat flavor created by star anise upon slow cooking. Although trans-anethole contributes a licorice-like flavor to the dishes, the new sulfur-containing compounds derived from trans-anethole that form during prolonged cooking are the secret to the unique flavor of dishes seasoned with star anise.

Home cooks should be aware that Chinese star anise is quite different from Japanese star anise (*Illicium anisatum* Linn), which contains a highly toxic substance called *anisatin* that has acute neurological effects on humans, fish, and animals. Unfortunately, the two varieties are nearly indistinguishable from each other, and there have been reports of Chinese star anise being contaminated with the Japanese variety, also known as *shikimi*.

REFERENCE

Pu, X. "Thermal Reaction of Anisaldehyde in the Presence of L-Cysteine, a Model Reaction of Chinese Stew Meat Flavor Generation." Master of science thesis, Rutgers University, New Brunswick, NJ, 2014.

Nutra. "FDA Warning on Star Anise Teas." September 10, 2003, https://www.nutraingredients-usa.com/Article/2003/09/11/FDA-warning-on-star-anise-teas.

The desire to cook food with diverse ingredients and new flavors has moved out of the realm of top-name chefs and restaurants and into the home kitchen. I like to think that the late, greatly admired chef Anthony Bourdain played a significant role in fueling home cooks' interest in exotic foods of the world, as well as in the diverse human cultures that created these wonderful cuisines and flavors. Christopher Kimball's new venture, Milk Street, is helping to encourage this renaissance for the home cook by developing simple recipes with bold flavors that use ingredients and cooking techniques from around the world, including Europe, the Middle East, and Asia. Although science is not an obvious tool in developing these recipes, it still plays a very important role in developing the techniques used, from the simple, such as blooming spices (roasting them in oil or butter, known as *tadka* in Indian cooking), to the more complex, such as making great, tender pie dough using a foolproof process based on *tangzhong* (a Japanese method in which boiling water is added to a small amount of flour to rapidly gelatinize the starch).

Cooking Is the Perfect Embodiment of Art and Science

Simple cooking is a craft that requires significant knowledge and skill, while the finest form of cooking is said to be an art. Knowing how to season a dish is a skill, but Thomas Keller's exquisite Peach and Summer Onions at the French Laundry is a work of art.

Even with these examples, the meaning of cooking as an art is not entirely clear. In the early 1970s, artist and photographer Martha Rosler wrote a fictitious dialogue between French chef Julia Child and New York restaurant critic Craig Claiborne titled *The Art of Cooking*.

JULIA CHILD: *Craig, my dear, I've been thinking. We all know that cooking is an art. But I wonder: How did it get to be one? After all, most of what we think of as art hangs on the wall or sits out in a courtyard.*

CRAIG CLAIBORNE: *Yes, cooking is an ephemeral art; the painter, the sculptor, the musician may create enduring works, but even the most talented chef knows that his masterpieces will quickly disappear. A bite or two, a little gulp, and a beautiful work of thought and life is no more.*

JULIA CHILD: *I see. You're talking of masterpieces and talent and beautiful work, but isn't cooking primarily a matter of taste?*

CRAIG CLAIBORNE: *Of course! But isn't all art a matter of taste?*

JULIA CHILD: *Yes! But are they the same kind of taste?*

CRAIG CLAIBORNE: *Classic French cooking is a fine art, as surely as painting and sculpture are. Its great works, such as poulard a la Neva and filet de boeuf Richelieu, are masterpieces designed to enchant both the palate and the eye. And the classic French cuisine implies as well the careful blend of textures, colors, and flavors; it means sparkling crystal, gleaming silver, and immaculate napery. When all these come together, it is one of the glories of the civilized world.*

So there you have an interpretation of the art of cooking from a famous artist. But, you may say, great art can be created without science. True, but science can significantly enhance the art of cooking, as it can any form of art. Consider that about 2,000 years ago all painting was two-dimensional with no perspective. Then about 1000 C.E., the great Arab mathematician and physicist al-Haytham recognized mathematically how the human eye sees objects in three dimensions, and within several hundred years, great artists like Leonardo da Vinci and Albrecht Dürer were creating great art with a three-dimensional perspective. Or consider how chemists of that era learned how to make beautiful stable pigments that greatly contributed to the exquisite colors of paintings by the Dutch artist Johannes Vermeer or the more recent American luminist painter Fitz Henry Lane. In the same way, a chef who knows that the green pigment chlorophyll in green beans is unstable in acid, which chemically changes the chlorophyll into a dull olive green pigment, can add just a tiny pinch of baking soda to retain the beautiful bright green color of fresh beans during cooking. Or consider that a fat like butter transfers much less heat energy to food than wine does, a realization that enabled Keller to create the luscious texture of his butter-poached lobster. By knowing how the muscle fibers in meat begin to shrink and squeeze out moisture at certain temperatures, a chef can cook meat that is tender and juicy. And, finally, Andrea Geary, a test cook with America's Test Kitchen, masterfully adapted her knowledge of the crystal structure of fat (with a little help from me) to create chocolate brownies that were supermoist and chewy.

As a professional scientist, amateur artist, and enthusiastic cook, I truly believe cooking has evolved over the past 200 years to become the perfect embodiment of art and science. The future of modern cooking is moving toward an emphasis on fresh seasonal local ingredients that are lovingly prepared, using both science and art, to produce exquisitely beautiful food enhanced with bold exotic flavors, making it pleasing for both the palate and the eye. Simultaneously, science-based cooking methods such as sous vide are being employed to retain and even enhance the nutritional quality of food, as explained in the final two chapters of this book. It's like having your cake and eating it, too.

6 Cooking Science Catches Fire!

Cooking Science Goes Viral

I first met Harold McGee in January 2005 at a coffee shop on University Avenue in Palo Alto, California, where I asked him to autograph the second edition of his popular book *On Food and Cooking: The Science and Lore of the Kitchen*, published the previous year. For me, the book was very inspirational, as I was going to start working part-time as the science editor for America's Test Kitchen in March 2005. The original founder of the test kitchen, Christopher Kimball, had a sincere interest in cooking science and its application in the development of recipes that worked for home cooks, so I was really excited to start this new journey after retiring from the food ingredients industry. Over coffee, McGee confided that the first edition of *On Food and Cooking*, published in 1984, had not sold well but that the new, significantly revised edition was a very big success. The new edition has received a James Beard book award and an International Association of Culinary Professionals cookbook award, and it's hard to find a serious cook today who does not own a copy.

What happened in the 20 years between the first and second editions of the book that accounted for one being a disappointment and the other being a huge success? Simply put, home cooks and professionals became fascinated with cooking science. After the first edition of *On Food and Cooking*, *McGee* published a second book in 1990 called *The Curious Cook: More Kitchen Science and Lore*, which didn't let the public forget about

FIGURE 6.1

A beautiful, delicious, and healthy dish of pan roasted grouper and roasted carrots on a bed of legumes and whole grains, prepared by sustainable seafood chef Rick Moonen. Photograph by Sabin Orr.

cooking science. Clearly, McGee deserves the bulk of the credit for awakening the public to the joys of cooking science, and his 2004 book was a barometer of that awakening. But I believe the real interest that led to McGee's success started with the publication of *Cook's Illustrated* magazine by Christopher Kimball. Kimball had started publishing a cooking magazine called *Cook's Magazine* in 1979 and soon after sold it to Condé Nast Publications, which eventually stopped publication in 1989. Kimball then began publication of *Cook's Illustrated* in 1992.

The inaugural issue of *Cook's Illustrated* contained cooking science notes written by McGee on the need for scalding milk and the science of butter. It also included a full article on "The Way to Roast Chicken," for which McGee roasted more than a dozen chickens under a variety of conditions and temperatures to find the best method. His scientific approach to developing a recipe became the formula for success for *Cook's Illustrated*. That groundbreaking 1992 issue also contained a large anonymous sidebar on "The Science of Egg Whites." The cooking public loved it, and subscriptions grew twice as fast as for any other cooking magazine, reaching a total of 1.2 million subscribers by 2009. McGee's early contributions to the magazine were critical because they gave the science credibility. Even today there is far too much fabricated cooking science (Should I say made-up cooking science?) being published, especially on the internet. Be careful if you read a book or blog on the science of cooking that does not back up the "facts" with either valid kitchen tests or references to the source of this information in the scientific literature. Otherwise, you are relying on the author as the source, and who knows from where they may have gotten the science. Note that McGee has fifteen pages of references at the back of *On Food and Cooking*. Is it any wonder it is such a valuable resource?

The public's interest in science-based cooking was further fueled by the very popular television show *America's Test Kitchen*, launched in 2000 and now producing shows for its nineteenth season. On each program, artful animation that catches the audience's interest illustrates the science behind selected recipes. To be sure, other developments have bolstered the interest in cooking science, including the creation of the Food Network in 1993 and the airing of Alton Brown's very popular science-based show *Good Eats*, which started on a local Chicago television station in 1998 and moved to the Food Network the following year. And then there were the contributions of other authors such as biochemist Shirley Corriher, who wrote *Cookwise: The Hows and Whys of Successful Cooking* in 1997. After I personally pestered Kimball and executive editor Jack Bishop for 2 years to write a book on cooking science, the company finally published *The Science of Good*

Cooking in 2012. It rapidly became the second-highest-selling book for the company after appearing on the *New York Times* "Best Seller List" for 8 weeks. This was followed in 2016 by a second science book, *Cook's Science*, which was nominated for a James Beard book award. As the named coauthor of both books, I was responsible for researching and writing all of the science, based on a review of almost 350 scientific papers.

Is the current interest in cooking science sincere, or is it a passing fad? I believe the interest is sincere and will increase in coming years because, as we will learn in the next sections, cooking science teaches us how to prepare and cook healthy food. The Centers for Disease Control determined that 70 percent of all deaths in the United States in 2014 were due to chronic diseases and predicts that this proportion will increase to 75 percent by 2020 without changes in our diet and physical activity. The increasing incidence of diet-related chronic diseases, such as obesity, diabetes, cardiovascular diseases, dental diseases, and certain cancers, in both the developed and the developing worlds is drawing greater attention to what foods we eat and how we prepare and cook them to reduce our risk of developing these life-threatening diseases.

Science Brings a New Vision to Cooking Healthy Food

Thanks to research sponsored by the National Institutes of Health (NIH), recommendations promoted by the Food and Drug Administration (FDA), and dietary guidelines developed by the U.S. Department of Agriculture (USDA) and American Heart Association (AHA), as well as extensive academic research on nutrition and epidemiology, we now know quite a lot about the foods in our diet that are healthy and those that are unhealthy. For example, evidence compiled by NIH shows that foods rich in antioxidants—such as carotenoid pigments, polyphenols, and vitamins C (ascorbic acid) and E (tocopherols)—reduce the oxidative stress caused by the reactive free radicals formed in our body, thus reducing the risk of cancer, diabetes, and cardiovascular, Alzheimer's, Parkinson's, and eye diseases. On the flip side, many health organizations have compiled extensive information showing that some fats, but especially trans fats, significantly increase the risk of heart attacks and strokes, as well as causing inflammation, insulin resistance, and diabetes. Unfortunately, we know much less about how home cooking affects the nutrients in food. One of the reasons is that the levels of nutrients in fresh fruits and vegetables vary significantly between batches of the same crop depending on when, where, and how it was grown; when it was harvested; and how it was processed, shipped, and stored. As we will learn later in this chapter, cooking food can cause further significant changes in the levels of nutrients, both good and bad, depending on how it was cooked. Expanding our scientific knowledge of how cooking affects the nutritional quality of food will have a profound impact on human health, fueling increased interest in cooking science.

Since the late 1930s, research on how food processing, including both food service and home cooking practices, affects the approximately four dozen nutrients required for human nutrition has been steadily increasing. For a very thorough review of the research between the late 1930s and early 1970s, refer to *Nutritional Evaluation of Food Processing*, edited by Robert S. Harris and Endel Karmas (the second edition, published in 1975, is still in print). Although the extensive data are now quite dated, they are still valid because the chemistry of nutrients does not change and the relative impact of many forms of food processing on the nutrient content of food remains the same. Chapter 17, by Paul LaChance and John Erdman (both were associated with the Department of Food Science at Rutgers University), covers "Effects of Home Food Preparation Practices on Nutrient Content of Foods" and is clearly relevant for the home cook, while chapter 16, by Paul LaChance, on "Effects of Food Preparation Procedures on Nutrient Retention with Emphasis Upon Food Service Practices" is especially relevant for restaurant chefs. This book has great meaning for me because Robert Harris (1904–1983) was my mother's cousin and professor of nutritional biochemistry for 33 years at the Massachusetts Institute of Technology. He was also the senior editor for the thirty-one-volume series *Vitamins and Hormones*, published by Academic Press. Harris awakened my interest in nutrition, although it was soon superseded by my fascination with the structures of organic molecules and the beauty of the chemistry of steroids such as cholesterol, progesterone, and testosterone.

For those seeking a very readable, scientifically accurate book on healthy eating (with recipes), I strongly recommend the most recent version of *Eat, Drink, and Be Healthy: The Harvard Medical School Guide to Healthy Eating*, written by Walter C. Willett, MD, and published in 2017. Willett was chair of the Department of Nutrition at the Harvard T. H. Chan School of Public Health for 25 years and is one of the world's leading experts on nutrition. Finally, for those seeking a scholarly paper on the effects of cooking on nutrients in food, see "A Review of the Impact of Preparation and Cooking on the Nutritional Quality of Vegetables and Legumes," which I coauthored with Adriana Fabbri, a postdoctoral student from Sao Paulo, Brazil. See also "The Effect of Cooking on the Phytochemical Content of Vegetables," by Mariantonella Palermo, Nicoletta Pellegrini, and Vincenzo Fogliano.

Promoting the application of cooking science as a means of enhancing the nutritional quality of food has been an interest of mine for many years. In May 2013, I gave a lecture on "Pairing Cooking Science with Nutrition" at the Department of Nutrition of the Harvard T. H. Chan School of Public Health. The lecture was largely based on Seminars in Food Science, Technology, and Sustainability, the course I teach at the Department of Nutrition. In January 2014, I was invited to give essentially the same lecture as the Zagat guest chef at Google headquarters in Mountain View, California. At the time, the huge Google campus had about thirty-five independent cafeterias, most serving three free meals per day and nearly all focused on healthy food. My lecture was recorded for YouTube, should you want to watch it. Fortunately, the video does not show the unsettling drama that occurred just minutes before the lecture was recorded—when I had to reformat about a dozen of my slides because the Google projection equipment was not compatible with the slides prepared on my computer! The lecture was presented several more times, including at the 2015 Conference on Nutrition and Health, organized by Dr. Andrew Weil and held in Phoenix. Some of the following examples of how cooking affects the nutritional quality of food are adapted from these lectures.

Cooking Science Enhances the Nutritional Quality of Food

Food is composed of a very complex mixture of divergent components, which makes studying and understanding the science of cooking a significant challenge. The components of food can be divided into two broad categories, the macronutrients and micronutrients. The macronutrients consist of proteins, carbohydrates, fats, and water and comprise about 98 percent of food by weight, while the micronutrients consist of very small amounts of numerous vitamins, minerals, and, in the case of plant foods, substances called phytochemicals, many of which are bioactive and nutritionally important. Processes that occur during cooking primarily involve the macronutrients and affect the flavor, texture, and appearance of the food, as well as the digestibility of proteins and starch, while effects on the micronutrients have a greater impact on the nutritional quality of the food. Factors including cooking temperature, cooking time, pH, and cooking medium (such as water, steam, fat, or dry cooking) play a very important role. For example, increasing the volume of water when boiling vegetables increases the loss of water-soluble vitamins and minerals, so something as simple as how much water is used for boiling a vegetable will affect nutritional quality (a 5:1 ratio of water to food is commonly used).

TABLE 6.1
Maximum Cooking Loss of Vitamins in Food

Vitamin	Maximum Cooking Loss (%)
Vitamin A	40
Vitamin B6*	40
Vitamin B12*	10
Vitamin C*	100
Vitamin D	40
Vitamin K	5
Niacin*	75
Riboflavin*	75
Thiamin*	80
Folic acid*	100

Source: Harris and Karmas 1975.
* Water-soluble vitamins.

Vitamins are essential nutrients that our body cannot make or does not make in sufficient amounts, so we must obtain them from our diet. Table 6.1 shows the maximum vitamin losses that occur when food is cooked by a variety of methods. Not all methods of cooking result in such large losses, but the numbers do convey the relative stability of vitamins for a variety of cooking conditions. A word of caution is necessary when considering numbers like those shown in table 6.1: look for trends rather than putting too much weight on the absolute numbers because of the large variation that occurs in the composition of food. You will note that the greatest loss occurs with vitamin C (ascorbic acid) because it is water soluble, unstable to heat, and easily oxidized; thus it serves as the "canary in the coal mine" and is often used as a standard to compare the effects of different methods of cooking on the loss of nutrients in food. Fresh fruits and raw vegetables are therefore the best sources of vitamin C, which plays many important roles in the body. Vitamin C is involved in the synthesis of collagen, the primary protein in connective tissue that aids in healing wounds and reducing the onset of scurvy. As a reducing agent that functions as an antioxidant, vitamin C also plays an important role in reducing the oxidation of LDL cholesterol, thus providing protection against cardiovascular disease. Lack of vitamin C is best known as the cause of scurvy, indicated by the body's failure to produce sufficient collagen for bones and connective tissue, but the first sign of vitamin C deficiency is fatigue, a concern among many people.

Again, when considering the effect of cooking on the nutritional quality of food, it is very important to look for trends in the data rather than focusing on specific numbers because even the best-designed studies show significant variation in the numbers. For example, consider the important family of bioactive phytochemicals called *glucosinolates*, found in all cruciferous vegetables (about thirty different glucosinolates have been found in cruciferous vegetables). These glucosinolates and their breakdown products (called *isothiocyanates*) have been linked to a significant reduction in the risk of developing many types of cancers, including breast, stomach, and bladder cancers. However, cooking destroys nearly all of the isothiocyanates, which throws into question the identity of the cancer-fighting agents in cruciferous vegetables. Cooking studies have found that boiling these vegetables until tender (a valid end point) resulted in reported losses of 18–59 percent of the total glucosinolate content, with a mean loss of 37 percent. Ranges of reported cooking losses this wide are common in cooking studies. Furthermore, consider that the total glucosinolate content of fresh cruciferous vegetables varies five- to eightfold within different samples of the same vegetable due to differences in cultivars, soil fertility (including sulfur and mineral content), weather, cultivation practices, and postharvest storage conditions. Whether crops are grown conventionally or organically makes little difference. Coupled with the losses during cooking, this means the amount of beneficial phytochemicals in these foods can vary widely. There is one downside to consuming an excess of fresh cruciferous vegetables: one of the various isothiocyanates formed from the glucosinolates, aptly named progoitrin, spontaneously forms a compound that interferes with the uptake of iodine by the thyroid gland, resulting in goiter. But this occurs only with the consumption of fresh uncooked cruciferous vegetables.

Several years ago Adriana Fabbri, a postdoctoral student working with me at Harvard, decided to see if the glucosinolate content of broccoli had changed over the previous 25 years, perhaps due to changes in soil fertility (especially sulfur content) or fertilizer use. California produces 90 percent of the broccoli grown in the United States, and state records had shown that fertilizer use on broccoli had remained fairly constant (varying only a few percent) during this time, thus eliminating fertilizer use as a factor. A review was conducted of numerous studies analyzing the total glucosinolate content of broccoli grown in California (many different cultivars are grown) over a period of 25 years. The levels of total glucosinolates reported in the broccoli varied so widely that it was not possible even to see a clear trend based on when the broccoli was grown or to determine if changes in soil fertility had been a factor. This is why it is difficult to draw firm conclusions from a single set of numbers or even a single study; rather, any conclusions should be based on the results of many studies. In the field of nutrition, analyzing the trends reported in a large set of studies is called a *meta-analysis*. Yet we often read reports in the press or on the internet about a single study that has shown some dramatic effect of a food or nutrient on human health.

Not only is a very wide range of nutrient levels found in all crops, but also the stability of nutrients following harvest has a significant impact on the nutritional quality of fruits and vegetables. Table 6.2 shows the percentages of vitamin C lost when samples of five types of freshly harvested vegetables were stored under different conditions. Notice that the vegetables stored frozen (−4°F/−20°C) for 12 months lost less vitamin C (relative to the

(CONTINUED ON PAGE 124)

FIGHTING CANCER WITH CRUCIFEROUS VEGETABLES

The relationship between diet and cancer in humans has been recognized for more than 40 years. For example, a Western diet high in fat has been associated with an increased risk of colon cancer, while people living in China, Japan, and Korea who consume a low-fat diet are four to ten times less likely to develop colon or breast cancer than people living in the United States. Almost half of the cancers diagnosed in the United States occur in the lung, colon, rectum, breast, and prostate. Fortunately, there is substantial evidence that a diet high in fruits and vegetables reduces the risk of developing certain cancers due to the presence of bioactive molecules known as *chemopreventive agents*. One difficulty in finding clear relationships between diet and the suppression of cancers in humans arises from the wide variation in the levels of chemopreventive agents in fruits and vegetables caused by cultivation practices, soil composition, weather, environment (including water), fertilizer use, and postharvest treatment, including the impact of storage, preparation, and cooking.

For more than 30 years, cruciferous vegetables have been associated with a reduction in the risk of certain cancers, shown more recently to include breast and bladder cancers and possibly lung and even prostate cancers. Research has attributed the protective role of cruciferous vegetables to bioactive molecules formed within the plant cells when they are damaged by chewing and cutting. These are the same compounds that are responsible for the pungent taste and aroma of cruciferous vegetables. About thirty-six different cruciferous vegetables are consumed around the world, but in the United States, we tend to focus on one genus of this family known as *Brassica oleracea*, which includes kale, collard greens, Chinese broccoli, cabbage, savoy cabbage, Brussels sprouts, kohlrabi, broccoli, broccoli romanesco, cauliflower, and broccolini. Other commonly consumed forms are horseradish, bok choy, broccoli rabe, turnips, arugula, watercress, radishes, and wasabi. As we have learned elsewhere in this book, all cruciferous vegetables contain naturally occurring compounds (phytochemicals) called *glucosinolates*, which, upon cell damage, react with an enzyme called *myrosinase* to produce bioactive compounds chemically known as *isothiocyanates* (ITCs). In nature, these compounds act as deterrents to insect attack, so it is not surprising that they might produce biological effects in humans. The ITCs are believed to be the bioactive compounds primarily responsible for protection against various forms of cancer (indole 3-carbinol is another chemopreventive agent formed from glucosinolates), but as pointed out in this chapter, most of the ITCs are destroyed by cooking, which raises the question of the identity of the cancer-fighting agent(s) in cruciferous vegetables. Let's explore the possibilities.

Numerous animal and in vitro (isolated cell) studies have confirmed the role of ITCs in suppressing cancers by a number of mechanisms, including detoxifying carcinogens, protecting cells against DNA damage, inhibiting blood vessel formation in tumors, and inhibiting tumor cell migration (National Cancer Institute 2012). Even if the identity of the chemopreventive agents is uncertain, extensive studies strongly support the role of cruciferous vegetables in reducing the risk of cancer. But animal and cell studies are relatively easy to control, using specific doses of pure bioactive molecules that tend to give fairly clear results within a limited time frame. Comparable studies in humans are not possible except with very small, carefully controlled groups of subjects. Many of the epidemiological studies done with very large groups of people involve recalling what was eaten and when, how often, and in what quantity it was eaten. If these were the only limitations, given enough subjects over sufficient time, we could see fairly clear trends of diet versus health outcomes. Unfortunately, the wide variation in the level of bioactive molecules present or formed in fruits and vegetables, coupled with the impact of preparation and cooking methods, makes it very difficult to determine the actual amount of bioactive molecules consumed by the test subjects. For example, we know that the level of glucosinolates in a single crop like broccoli can vary five- to eightfold, and boiling can eliminate almost 60 percent of this variable amount. This means that some subjects who consumed the same number of servings of broccoli per week may have ingested many fewer or many more ITCs than other subjects. And this difference is compounded if one set of subjects steamed their broccoli and another set boiled theirs, as glucosinolates and ITCs are water soluble and easily leached into cooking water, while steaming has been shown to increase the level of ITCs.

Relatively few studies have been conducted comparing the amounts of ITCs consumed in raw versus cooked cruciferous vegetables. One such study conducted at the American Health Foundation (Getahun and Chung 1999) compared the amount of ITCs consumed in raw versus boiled watercress and then analyzed the amount of ITCs excreted in the urine. Subjects eating raw watercress not only consumed about 4.6 times more ITCs than those eating cooked watercress from the same batch but also excreted about 10 times more ITCs, indicating that cooking watercress in water significantly reduces the level of ingested ITCs both by leaching out the glycosinolates and ITCs and by deactivating the myrosinase enzyme. So even though subjects in a large study consumed the same number of servings of a cruciferous vegetable over a period of time, the amount of bioactive molecules ingested may have been very different for each subject depending on the source of the vegetable and how it was prepared.

The watercress study resulted in another interesting observation. Boiling the watercress for 3 minutes completely destroyed the myrosinase enzyme, yet some ITCs were still found in the urine of subjects who consumed the cooked vegetable, although much less than in that of subjects who consumed raw watercress. According to the authors, this was the first study to show that intact glucosinolates in a cooked cruciferous vegetable can be converted to isothiocyanates in humans. So cooking cruciferous vegetables may not eliminate all beneficial bioactive molecules. Furthermore, incubating cooked watercress juice containing glucosinolates, but no ITCs, with human feces produced a significant amount of ITCs; this indicates that some of the bacteria in feces from the gut microflora may possess myrosinase-like activity, making them another source of ITCs. So even though cooking destroys most of the ITCs formed from glucosinolates, it does not rule them out as the cancer-fighting agent formed in cruciferous vegetables because they are produced in multiple ways.

Taken as a whole, in vitro, animal, and human studies confirm that cruciferous vegetables help to fight certain cancers, especially bladder and breast cancers.

REFERENCES

Getahun, S. M., and F.-L. Chung. "Conversion of Glucosinolates to Isothiocyanates in Humans After Ingestion of Cooked Watercress." *Cancer Epidemiology, Biomarkers and Prevention* 8 (1999): 447–451.

Murillo, G., and R. G. Mehta. "Cruciferous Vegetables and Cancer Prevention." *Nutrition and Cancer* 41, nos. 1 & 2 (2001): 17–28.

National Cancer Institute. *Cruciferous Vegetables and Cancer Prevention*. Bethesda, MD: National Cancer Institute, 2012.

Willett, W., and P. J. Skerett. *Eat, Drink, and Be Healthy: The Harvard Medical School Guide to Healthy Eating*. New York: Free Press, 2017.

TABLE 6.2
Loss of Vitamin C (as % of Dry Weight) Due to Storage

Food	Fresh 68°F (20°C) 7 Days	Fresh 39°F (4°C) 7 Days	Frozen −4°F (−20°C) 12 Months
Broccoli	−56	0	−10
Carrots	−27	−10	0
Green beans	−55	−77	−20
Green peas	−60	−15	−10
Spinach	−100	−75	−30

freshly harvested vegetables) than did those stored for 1 week at refrigerator temperature (39°F/4°C) or at room temperature (68°F/20°C). This is the case with many types of frozen vegetables because they are almost always rapidly blanched and flash frozen within a few hours after harvest. The vitamin C content of spinach and green beans is very susceptible to degradation because fresh vegetables contain active enzymes that catalyze the oxidation of vitamin C during storage. Blanching before freezing deactivates the enzymes. In addition to vitamin C, other fairly unstable vitamins in vegetables are folic acid and thiamin. You can imagine that the loss of less stable nutrients like vitamin C can be quite significant in fruits and vegetables that are shipped across the United States or from foreign countries and then stored at a distribution center before finally arriving at the supermarket, where many vegetables are kept moist, but are not refrigerated, for several days. In many cases, these so-called fresh vegetables should be stored in the refrigerator as soon as they are brought home because they lose fewer nutrients there than at room temperature. For the past 10–20 years, controlled atmosphere storage and modified atmosphere packaging methods have been utilized to prolong the shelf life and improve the nutrient stability of fresh fruits and vegetables. In both cases, the level of oxygen in the package or storage chamber is reduced from about 22 percent, the level in the atmosphere, to about 2.5 percent, while the level of carbon dioxide is increased from about 0.04 percent, the level in the atmosphere, to 2.5 percent. The difference is made up by adding more nitrogen. Increasing carbon dioxide and decreasing oxygen greatly retards the respiration and spoilage of fresh fruits and vegetables.

Cooking further exacerbates the destruction of some (but not all) nutrients in food, but the methods and conditions of cooking greatly influence the extent of loss or gain. Tables 6.3, 6.4, and 6.5 illustrate very effectively how three common cooking methods

TABLE 6.3

Loss or Gain of Nutrients in Cooked Broccoli

Nutrient	Raw	Boiled	Steamed	Fried
Total carotenoids	28*	+32%	+19%	−67%
Total phenols	100*	−73%	−38%	−60%
Vitamin C	847*	−48%	−32%	−87%
Total glucosinolates	71†	−59 %	+30%	−84%
Cooking time		8 mins.	13 mins.	3 mins.

*Numbers are expressed in milligrams of nutrient per 100 grams of food on a dry weight basis.
†Number expressed in micromoles (molecular weight) per gram of food on a dry weight basis.
Percentages are expressed as increase or decrease of nutrient on a dry weight basis.

TABLE 6.4

Loss or Gain of Nutrients in Cooked Carrots

Nutrient	Raw	Boiled	Steamed	Fried
Total carotenoids	118*	+14%	−6%	−13%
Total phenols	70*	−100%	−43%	−31%
Vitamin C	31*	−10%	−39%	−100%
Cooking time		25 mins.	30 mins.	8 mins.

*Numbers are expressed in milligrams of nutrient per 100 grams of food on a dry weight basis; percentages are expressed as increase or decrease of nutrient on a dry weight basis.

TABLE 6.5

Loss or Gain of Nutrients in Cooked Zucchini

Nutrient	Raw	Boiled	Steamed	Fried
Total carotenoids	50*	−4%	−22%	−35%
Total phenols	59*	−70%	−41%	−63%
Vitamin C	194*	−4%	−14%	−14%
Cooking time		15 mins.	24 mins.	4 mins.

*Numbers are expressed in milligrams of nutrient per 100 grams of food on a dry weight basis; percentages are expressed as increase or decrease of nutrient on a dry weight basis.

FIGURE 6.2

A texture analyzer similar to an earlier model used by the author for his research at Framingham State University. Reproduced with permission from Brookfield Engineering.

affect the percentages of specific nutrients lost or gained for three vegetables. Out of many similar studies that I could cite, I have selected this one from the Universities of Parma and Napoli because the design and breadth of the study are so well thought out. You may be thinking that I have cautioned against relying on a single study, but the results of this study are so consistent with many others that I am comfortable using it as an example.

Most studies of this nature measure the loss of nutrients after cooking the vegetables for a specific length of time, such as 10 minutes. A specific cooking time is useful for comparing the loss of one nutrient with that of another or one vegetable with another. But at home, we cook vegetables until they reach a certain tenderness, which is how this study was conducted, making it more representative of home cooking than a laboratory study. The researchers determined how long it took to cook carrots, zucchini, and broccoli to the desired tenderness, as determined by a trained taste panel. The end points were then quantified by measuring the tenderness with a very sensitive laboratory instrument called a *texture analyzer*, which measures the force necessary to penetrate a sample of food (figure 6.2). The forces needed to penetrate both the raw and the cooked vegetables were then used to calculate the *percent softening* after cooking each vegetable to the desired tenderness under carefully controlled conditions. All tenderness measurements were performed at 122°F (50°C), typical of the temperature at which the vegetables would be consumed. Measurements were then made of the levels of specific nutrients in samples of the raw and cooked vegetables, with each sample coming from the same batch of the vegetable being tested to ensure there was no difference in the nutrient content between samples of the same raw vegetable. Three samples of each vegetable were cooked by each method, and the nutrient content of each sample was measured ten times to ensure reproducibility. The nutrient contents were measured for total carotenoid pigments (the yellow, orange, and red pigments in vegetables; beta-carotene is the precursor of vitamin A), total polyphenols, vitamin C (ascorbic acid), and total glucosinolates. The first three nutrients function as antioxidants in food, while the glucosinolates and their breakdown products, the isothiocyanates, are bioactive phytochemicals. A diet rich in these four nutrients reduces the risk of developing certain cancers, as well as cardiovascular and other degenerative diseases. Notice that only table 6.3 contains glucosinolates because these phytochemicals are present only in cruciferous vegetables, like broccoli.

Before venturing into an explanation of all these numbers, let me mention that these studies were conducted in Italy, where the desired end points of tenderness selected by the taste panel were representative of Italian tastes and may not be the same for cooks in other countries. Based on the cooking times, you can decide if you would cook the vegetables for a longer or a shorter time to achieve the tenderness you prefer (Nicoletta Pellegrini, one of the authors, kindly provided me the cooking times). The boiling time was based on adding the vegetable to already boiling water (using a 5:1 ratio of water weight to vegetable weight), steaming was performed at atmospheric pressure where the temperature of the steam is the same as that of boiling water (212°F/100°C), and deep-fat frying was performed in 9.3 cups (2.2 liters) of peanut oil at 338°F (170°C). In addition, the vegetables were prepared for cooking as they traditionally are in Italy, which may differ from the methods of preparation in other countries (such as the method of cleaning, chopping, and the size of the chopped vegetables). Even given these differences, it is obvious that cooking vegetables has a significant impact on their nutritional quality.

So what trends can we infer from these carefully designed and executed experiments? First, of the three cooking methods, steaming is the least destructive of the nutrients, while high-temperature deep-fat frying is the most destructive (particularly for vitamin C in broccoli and carrots). Although steaming took the longest time to cook vegetables to the desired tenderness, it generally did not extract water-soluble nutrients (such as vitamin C, polyphenols, and glucosinolates) the way boiling did, while frying involved the highest temperature and extracted oil-soluble nutrients like the carotenoids. The one drawback with steaming is that the lengthy cooking time promoted oxidation of light-sensitive nutrients such as the polyphenols, as we see for zucchini. The other important observation is that cooking actually increased the availability of some nutrients, especially the carotenoids in carrots and broccoli (the chlorophyll that makes broccoli green also masks the modest amount of carotenoids that are present). Steaming also increased the availability of glucosinolates in broccoli, presumably because they were released from the cell walls by the heat.

That cooking increases carotenoids has been well documented for lycopene, the major red carotenoid pigment in tomatoes. Studies have shown that the amount of fat-soluble lycopene absorbed into the blood is almost four times higher from cooked tomato products (such as sauce and paste) than from fresh tomatoes and that cooking tomatoes in an oil such as olive oil increases the amount of lycopene absorbed into the blood by 80 percent. This is very important because epidemiological studies have shown that consuming two to three servings of tomato sauce per week reduces the risk of all forms of prostate cancer by 35 percent and that of the most advanced forms by 50 percent. In fresh tomatoes, lycopene is bound to proteins, which limits its absorption into the body, while cooking releases the lycopene from the proteins, so it is more readily absorbed with fats and oils.

Carotenoids are not the only nutrients whose availability and absorption are positively affected by cooking, as shown by research conducted by the USDA. Researchers there studied how cooking a variety of vegetables—including kale, broccoli, cabbage, collard greens, mustard greens, Brussels sprouts, spinach, and bell peppers—influenced the binding of bile acids in the intestine. Bile acids are responsible for solubilizing fats, oils, and oil-soluble vitamins, which allows them to be absorbed into the body from the intestine. Bile acids are steroids synthesized in the liver from cholesterol, a process that consumes at least half of the approximately 800 milligrams of cholesterol produced each day in the body. However, about 95 percent of bile acids are reabsorbed back into the body and recycled for further use in the intestine, thus limiting the amount of cholesterol used for synthesis. Cooking the vegetables listed by boiling (10–14 minutes), steaming (10–20 minutes), and sautéing (15–20 minutes) increased their ability to bind and eliminate bile acids from the body compared to that of raw vegetables. The net result of eating these cooked vegetables is that more bile acids are excreted from the body (rather than being recycled); thus more cholesterol is consumed and its level in the blood reduced. The effect of cooked vegetables was compared with that of *cholestyramine*, a drug approved by the FDA for the reduction of bile acids and cholesterol. The results of the in vitro laboratory study showed that 100 grams of cooked vegetables were 4–14 percent as effective as 100 milligrams of the drug. This may not sound like much, but consider that these levels are the same as or slightly higher than the bile acid binding values of oat bran and ready-to-eat oat bran cereals, which have been approved by the FDA for a heart healthy label because they lower cholesterol. The most effective vegetables and cooking methods were sautéed or steamed kale and mustard greens, followed by steamed or sautéed collard greens, and then sautéed broccoli, cabbage, and bell pepper. Evidence suggests that cooking converted some of the insoluble dietary fiber in the vegetables to soluble fiber that was responsible for binding the bile acids. So next time you are looking for a different vegetable to cook, try sautéing peppery mustard greens and a little chopped garlic in olive oil and then briefly simmering them with a few tablespoons of chicken stock or bouillon until they are tender. If you don't like the strong peppery taste, try blanching the greens in hot water for about 30 seconds before sautéing to deactivate the myrosinase enzyme that produces the pungent taste (see chapter 5).

RECIPE 6.1 Tomato Sauce With Red Bell Peppers

INGREDIENTS:

3 Tbsp. extra virgin olive oil

1 medium red bell pepper (or ½ large red bell pepper), cut into thin strips

1 carrot, peeled and chopped into small pieces

1 medium to large yellow onion, chopped into very small pieces

4 garlic cloves, finely chopped

3 ounces tomato paste (half of a 6-ounce can)

1 28-ounce can ground peeled tomatoes

1 tsp. sugar

1 Tbsp. dried oregano

1 Tbsp. dried basil

***Yield*: 6–8 servings with pasta**

This sauce is a rich source of healthy orange- and red-colored pigments called carotenoids, especially lycopene from tomatoes, capsanthin from red bell peppers (and also found in paprika), and beta-carotene from carrots. Carotenoids function as antioxidants and are associated with a reduced risk of cardiovascular disease and some cancers, such as prostate and lung cancers, as well as age-related macular degeneration (a few of the carotenoids, such as beta-carotene, serve as precursors of vitamin A). There is some evidence they may also help to maintain cognitive health. Carotenoids in foods tend to be bound to proteins, which reduces absorption of the carotenoids into the body from foods such as fresh tomatoes. Cooking releases carotenoids from the proteins, making them more accessible for absorption. All carotenoids are oil soluble, so preparing tomato sauce with olive oil aids in their absorption from the small intestine and thus increases their levels in the blood. Here the prolonged heating not only releases more of the carotenoids into the oil, making them available for absorption, but also helps to oxidize some of the carotenoids to produce a more complex flavor.

My mother first made this recipe and passed it on to my older sister and eventually to my wife, Christine, and then to me. It differs from other basic tomato sauces by calling for a significant amount of red bell pepper, which not only enhances the flavor but also provides another rich source of red carotenoids. The recipe makes enough sauce for six to eight servings of pasta. Because the sauce is made with acidic tomato puree, it has a low pH (<4.6) that retards the growth of bacteria and mold, so any leftover sauce will keep in the refrigerator for several weeks. Leftover sauce is also great for making a bean soup, as well as a topping for pizza.

If you are like my wife and want to restrict your consumption of pasta, then serve the sauce over diagonally sliced zucchini sautéed with garlic and olive oil, which she prefers to pasta. Read box 7.1, Winter Vegetable—The Science of Pasta, to learn why its reputation as a high-starch, high-glycemic-index food is not justified.

DIRECTIONS:

Add the olive oil, red bell pepper, and carrot to a 2-quart saucepan, and cook on medium heat, stirring occasionally, for about 5 minutes to soften and brown some of the edges of the red bell pepper and carrot. Add the onion, and continue cooking on medium heat, stirring occasionally, for another 4 minutes, until the onion is soft and perhaps even a little brown on the edges. Don't worry if the vegetables become too dark, as you really can't overcook them for this recipe. Add the garlic, and continue to cook for another minute, stirring frequently to prevent burning the garlic, which makes it bitter. Add the tomato paste, and mix with the vegetables to help draw out the carotenoids from the paste (notice how red the oil has become). Then add the tomatoes, and mix thoroughly. Add about ½ cup of water, and even more if the sauce is too thick, and reduce the heat.

Simmer the mixture for about 45 minutes, adding more water if needed to maintain the same volume and consistency so the sauce does not burp out of the pan. Stir the sauce as necessary to prevent burning. Add the sugar to offset the acidity of the tomatoes, season with the oregano and basil, mix thoroughly, and continue to simmer on low, stirring occasionally, for about 1 hour (tasting and smelling as you go). Keep adding water as necessary to prevent the sauce from becoming too thick or burning. Taste, smell, and add salt, pepper, and more sugar as desired. Thin the sauce with more water if it is too thick for your taste.

Other methods of cooking, including microwaving, pressure cooking, stir-frying, and oven roasting, have been studied for their effects on the nutritional quality of food. In general, heating in a microwave oven for less than 3 minutes, such as when warming food, does little to affect the nutrients, including vitamin C and the glucosinolates. But using longer heating times (over 15 minutes), as when fully cooking vegetables and meats, can have a significant impact on nutrients because the energy in a microwave oven is high enough to destroy many heat-sensitive molecules. For vegetables, you would be better off steaming them even if it takes a little longer.

Pressure cooking has somewhat less destructive impact on nutrients compared with boiling in water at atmospheric pressure. Even though pressure cookers heat the water to 250°F (121°C) under 1 atmosphere (15 pounds per square inch) of pressure, the cooking time is significantly less, so cooking food to an end point based on texture (rather than time) in a pressure cooker takes much less time than cooking in boiling water and the food suffers about 15–20 percent less destruction of nutrients. But remember that any food cooked in water will lose water-soluble vitamins and minerals into the cooking water. To avoid the loss of valuable nutrients when braising meats and vegetables, the braising liquid should be used to prepare soup, sauce, or gravy.

Stir-frying cooks foods very rapidly. However, because it requires a very high temperature and small pieces of food (with high surface area), it actually destroys most of the vitamin C and polyphenols.

The effect of oven roasting is best judged by its impact on thiamin (also known as vitamin B1). Thiamin is an important vitamin; it is involved in the metabolic pathway of respiration, providing an important source of energy for the body, and is also believed to play a role in nerve function. Meats are good sources of thiamin, so thiamin is a good indicator of the impact of oven roasting on nutrient retention. For some unexplained reason, pork contains about ten times more thiamin than beef, lamb, poultry, or fish. Because thiamin is water soluble and unstable to heat, most of its loss occurs when food is boiled or when cooked meat produces drippings. As an example, early studies showed about a 42 percent loss of thiamin from chicken roasted to an internal temperature of 190°F (88°C), some of which was presumably due to loss of moisture. During roasting, meat such as beef and poultry can lose as much as 25 percent of its moisture as the muscle fibers shrink due to heat. Always allow roasted meat to rest at least 15 minutes before slicing to provide time for the juices to be reabsorbed by the muscle tissue; this avoids the loss of thiamin and other water-soluble nutrients through drip loss.

RECIPE 6.2 Delicious, Healthy Mashed Cauliflower

INGREDIENTS:

4 cups 1½-inch pieces cauliflower

1 garlic clove, smashed

½ tsp. salt

¼ cup (4 Tbsp.) crème fraîche

NOTE:

Other seasonings, such as ground pepper or hot sauce, can be added if desired.

***Yield*: 4 servings**

This recipe uses a very simple cooking method, based on the science discussed near the end of chapter 5, that retains the flavor and nutrients in mashed cauliflower, making it a more nutritious substitute for mashed potatoes. Rather than being full of rapidly digested starch, cauliflower is high in dietary fiber and much lower in starch than potatoes. Also, because none of the cooking water is removed, all of the nutrients, such as the glucosinolates, isothiocyanates, water-soluble vitamins, fiber, and minerals, leached from the cauliflower will still be present. The cooking time is important. Cauliflower must be cooked long enough to retard the initial rapid formation of sulfury, pungent isothiocyanates and to allow the conversion of any that do form to nutty-tasting disulfides and trisulfides—but not so long as to allow the latter compounds to evaporate, leaving a bland, dull cauliflower mash.

For a flavorful, healthy meal, spread the cauliflower puree evenly on plates, top with a protein source such as oven-baked fish or grilled chicken and accompany with a leafy green vegetable or salad. Cooking other vegetables in this way in the minimum amount of water will result in a concentrate you can use without straining to make soups and sauces.

DIRECTIONS:

Place the cauliflower in a 2-quart saucepan, and add just enough water to cover only about half of the pieces. Add the garlic and salt. Simmer the cauliflower uncovered for at least 20 minutes (up to 25 minutes), until it is very soft and tastes and smells nutty and sweet rather than harsh and sulfury. Most of the water should have been absorbed by the cauliflower or evaporated, so there is no need to drain excess water from the cauliflower. Next add the crème fraîche, and thoroughly puree the cauliflower with a hand-held blender until it is smooth and creamy.

Do you remember when fat was the villain and reduced-fat food became the rage? Food companies scrambled to reduce the fat content of almost every food that contained fat, especially saturated fat, which was linked to increased levels of cholesterol in the blood and to heart disease. This effort occurred in the mid- to late 1980s and continued to increase through the 1990s. Jumping on the bandwagon in 1987, the National Pork Board launched an advertising campaign to promote "Pork. The Other White Meat." The campaign presented pork as more like healthier, lower-fat chicken than red meat like beef because beef was higher in saturated fat than chicken. Well before the start of the advertising campaign, pork breeders were taking steps to produce leaner pigs. The meat-eating public was well aware of the concerns about eating red meat, resulting in chicken consumption overtaking that of beef in 1992. The pork producers wanted pork to be perceived as more like chicken.

The U.S. Department of Agriculture (USDA) has always classified pork as a red meat because of its similarity to other red meat in fat content and composition. Newer generations of leaner pork may have less fat than older generations, but the composition of the fat is still about the same, largely due to the cereal grain–based diet fed to pigs and cattle since the 1970s. Pigs with less fat might be a healthier form of red meat, yet unlike the case with chicken, the per capita consumption of pork has remained almost constant since 1970, approaching 50 pounds per person per year in the United States. Could it be that leaner, paler pork is less appealing to the consumer?

Leaner pork is less flavorful and less moist, and it requires careful attention to cooking conditions to produce meat that is not as tough and dry as shoe leather. Is it any wonder sales have not kept pace with those of chicken? Until just a few years ago, the USDA recommended cooking whole cuts of pork to 160°F (71°C) to avoid any illness due to trichinosis parasites. But moving hogs indoors and feeding them a well-controlled, safe diet has lowered the incidence of trichinosis to about two cases per year in the entire U.S. population. In 2011, the USDA changed its recommendation to cook whole cuts of pork to 145°F (63°C), followed by a 3-minute rest. Ground pork must still be cooked to 160°F like ground beef, but a whole cut of pork can be cooked so it is slightly pink inside and much more tender and moist. Although this recommendation is moving in the right direction, it does not result in more flavorful meat.

Buying pork in the supermarket is a guessing game. How does one know if a cut of pork will be flavorful, moist, and tender? Is it even possible to buy pork with these attributes, and if so, how does one identify it? With a little knowledge of the science of pork, it is actually quite easy. It is based on the color of the fresh meat. To understand the science, let's step back and look at how the quality of pork is categorized and how pork is produced. The three standards for pork quality are the ideal standard, RFN (stands for reddish-pink, firm, and nonexudative, or no exuded water in the package), and the two lower standards, PSE (pale, soft, and exudate) and DFD (dark, firm, and dry). PSE is inferior in texture (mushy), flavor, and moistness. DFD is actually of exceptional quality with excellent water-holding capacity and tenderness. But because of its dark red color, consumers wrongly believe it is meat from older animals or lacks freshness. About 25 percent of all pork falls into the less desirable PSE and DFD categories.

The three critical stages in pork production that affect the quality of the meat occur before, during, and after slaughter of the animal. Nowadays slaughtering animals and birds for food consumption is referred as *harvesting*. Not surprisingly, the amount of stress experienced by the pig before and during slaughter has a profound effect on the quality of the meat. When animals are stressed, they produce higher levels of lactic acid, which lowers the pH of the meat. The pH of muscle in live animals is neutral, or pH 7. When excessive lactic acid is produced during stress, the pH drops significantly—to as low as 5.2–5.5 (moderately acidic). The pH of the muscle tissue determines if the meat will be tender and juicy or tough and dry. It also has a direct bearing on the flavor of the meat. The pH of pork is measured 45 minutes after slaughter and typically should be close to pH 6.2—and preferably closer to pH 6.5. If the pH is lower—for example, pH 5.7—the muscle fibers will tighten and hold less moisture when the meat is cooked. But, most importantly, the enzymes called *calpains* that break down the muscle proteins and make the meat tender and flavorful are most active close to pH 7. These enzymes continue to tenderize meat after the animal has been slaughtered. Meat that has a pH of 6.5 will be much more tender and flavorful than meat that has a pH of 5.7 or lower.

But, of course, we can't bring a pH meter to the supermarket when we buy pork. Fortunately, there is a simple way to tell the pH of the meat, and that is by its color. The pigment that colors all red meat is a protein called *myoglobin*. It is found in all muscle cells and is there to store the oxygen that muscles need for action. *Hemoglobin*, another red-colored protein, is found in the blood, where it transports oxygen from the lungs to the muscle cells. Hemoglobin is actually composed of four myoglobin molecules linked together. But hemoglobin is too large to enter muscle cells. In order to transport the oxygen into the cells, each hemoglobin molecule must dissociate into four myoglobin molecules, which then ferry the oxygen into the muscle cells. The intensity of the red color of myoglobin is dependent on the pH. The higher the pH, the redder the myoglobin. Thus we can simply look at the pork and gauge its pH. Darker red–colored meat means higher pH. The darker red color means the pork will be more tender, moist, and flavorful. Unfortunately, consumers who shun dark, firm, and dry (DFD) pork because they think it is not fresh are rejecting the best cuts of pork. So next time you are in the supermarket, look for cuts of pork that are dark and well marbled with fat and that contain no free water in the package. Pass up those pale, mushy-looking cuts because they will be tough, flavorless, and dry.

Look at the port cutlets in figure 1. Which would you prefer?

Some of the top breeders of pigs are now taking great care to raise and slaughter their animals with the minimum amount of stress, producing meat that is dark, flavorful, and tender. Pork from these pedigreed pigs is expensive but worth it—not only because the quality of the meat is higher but also because the animals have been treated more humanely. Next time you are in the supermarket, notice how dark the pork has gotten. Perhaps the National Pork Board will change its slogan to "Pork. The Other Red Meat"!

FIGURE 1

A comparison of pork cutlets with different colors due to the pH of the meat. The darker-colored pork (*a*) has a higher pH than the lighter-colored pork (*b*). Photographs from FeaturePics.

REFERENCES

Buege, D. *Variation in Pork Lean Quality*. Clive, IA: U.S. Pork Center of Excellence, 2003 (originally published as a National Pork Board /American Meat Science Association fact sheet).

Warriss, P. D. *Meat Science: An Introductory Text*, 102–204. Wallingford, Oxfordshire, UK: CABI, 2000.

Whole grains and legumes are also good sources of thiamin, much of which resides in the bran layer or outer coating. Again because it is water soluble, almost 50 percent of the thiamin in legumes is removed when beans are soaked and then cooked in water. Other water-soluble vitamins such as riboflavin and niacin are also leached from beans during soaking and cooking (about 50 percent and 70 percent, respectively). Polishing rice to remove the outer hull and germ results in thiamin reduction. Rice that has been parboiled in preparation for polishing loses significantly less thiamin because the vitamin diffuses into the endosperm of the rice during the parboiling step. Parboiling is not quite what it sounds like; the rice is steeped in water and then steamed and dried rather than being simply boiled in water, which would result in much greater loss of thiamin. Finally, one other significant loss of thiamin occurs in baked bread; as much as 30 percent of the thiamin in wheat flour may be lost during baking.

Sous vide cooking is moving from the restaurant into the home kitchen with the creation of more affordable cooking devices. The advantages of this method are twofold when it comes to enhancing the nutritional quality of food. First, cooking food at precisely controlled lower temperatures in vacuum-sealed pouches reduces the loss of nutrients by both excessive heat (which affects thiamin, folic acid, and B vitamins) and oxidation (which affects vitamin C and phytochemicals). Second, all juices are retained within the pouch so water-soluble vitamins, minerals, and phytochemicals are not lost, especially if the juices are used to make a sauce or flavoring. These effects are true for meat, poultry, fish, and vegetables. Not only does sous vide cooking produce melt-in-your mouth textures, but also it protects the true flavor and nutritional quality of the food. What more could you ask for?

7 The Good, the Bad, and the Future of Cooking Science

Good Carbohydrates and Bad Carbohydrates

When choosing a healthy diet, one of the most challenging decisions is that between good and bad carbohydrates, and good and bad fats, all of which are macronutrients that we consume in much larger amounts than micronutrients. Dietary carbohydrates range from simple sugars (such as fructose and glucose, both of which are *monosaccharides*, and sucrose, a *disaccharide*), to medium-sized molecules called *oligosaccharides* that are composed of three to ten simple sugars linked together (including raffinose, stachyose, and inulin), to very large *polysaccharides* composed of hundreds or thousands of interconnected sugar molecules (including starch, cellulose, hemicelluloses, pectin, and beta-glucans). Dietary fats run the gamut from solid saturated fats such as *tallow* (beef fat) and *lard* (pork fat) to a wide range of liquid *mono- and polyunsaturated* oils, including olive, soybean, and grape seed oils. Today the consumption of sugars added to processed and prepared foods (called *added sugars* to differentiate them from the natural sugars in fruits and vegetables) has become a major health issue, as it increases the risk of obesity, diabetes, heart disease, and liver disease. Added sugar increases the level of glucose in the blood, causing increased secretion of insulin, which, among its many physiological effects, causes the body to store excess food calories as fat in adipose tissue and creates the potential for developing insulin resistance, obesity, diabetes, and cardiovascular disease. Largely due to industry pressure, the Food and Drug Administration (FDA) recently postponed the requirement to label the amount of added sugars in processed food until 2020. The American Heart Association

FIGURE 7.1

Painting on silk, by Yan Ciyu, mid- to late twelfth century. The painting depicts the mystery of what lies ahead in the future, which we cannot see but only imagine.
Reproduced with permission of the Freer/Sackler Art Gallery of the Smithsonian Institution.

recommends limiting added sugars to 6 teaspoons per day for women, 9 teaspoons per day for men, and 3–6 teaspoons per day for children. In 2017, the average American consumed 22 teaspoons per day of added sugars, or more than 66 pounds per year. The major sources of added sugars in the American diet are soda and sports drinks, which provide almost half of all the added sugars consumed by children and adults. A single 12-ounce can of some sweetened drinks contains as much as 11 teaspoons of added sugars.

Many consumers have come to recognize that the presence of oligosaccharides in food is fairly widespread, as evidenced by the recent interest in the FODMAP diet developed by Monash University in Australia to treat bloating, gas, heartburn, and pain in the gastrointestinal system. FODMAP stands for *fermentable oligosaccharides, disaccharides, monosaccharides, and polyols (sugar alcohols)*. Most of the symptoms are caused by the oligosaccharides, which are very poorly digested in the small intestine and reach the large intestine where certain bacteria eagerly ferment them to produce gas and the other unpleasant responses. So we can put added sugars and possibly oligosaccharides in the basket of "bad" carbohydrates, although the oligosaccharides do not affect the level of glucose in the blood or insulin secretion and therefore are not involved in fomenting the harmful diseases caused by added sugars.

Beans are an inexpensive food that is considered to be very healthy because it is high in protein, minerals, and dietary fiber. In the fall of 2017, a student from Mexico City taking my class at Harvard on Seminars in Food Science, Technology, and Sustainability told me that in the past 25 years the consumption of beans in Mexico has declined by about 50 percent, apparently because they are thought of as food for the poor but also perhaps because they can cause gastrointestinal problems. Some of that fiber in beans is due to the small oligosaccharides *raffinose* (three sugar molecules linked together) and *stachyose* (four sugar molecules linked together), which cause the gas and flatulence associated with beans when fermented in the large intestine. The U.S. Department of Agriculture (USDA) has shown that heating dry beans to the boiling point in a large volume of water, allowing them to rest (hydrate) off the heat for 1 hour (as an alternative to soaking overnight), and thoroughly rinsing them eliminates 40–50 percent of these two troublesome oligosaccharides, thus reducing the tendency of cooked beans to produce gas. The hydrated beans can be immediately cooked in soups, stews, and chilies, or they can be frozen for use later. Canned beans are a convenient alternative to dry beans, but they are very high in salt, added primarily to maintain the texture of the beans in the can. One study has shown that thoroughly rinsing canned beans removes about 40 percent of the salt (and sodium), so it is always best to rinse canned beans before cooking. Rinsing also removes some of the oligosaccharides dissolved in the residual canning liquid.

RECIPE 7.1 White Bean and Roasted Chicken Chili

INGREDIENTS:

2–2½ total ounces of jalapeño and serrano or poblano chilies

1 Tbsp. extra virgin olive oil

1 medium onion, coarsely chopped

1 garlic clove, finely chopped

2½ cups chicken stock (made as above or purchased)

6 ounces chopped white meat from leftover roasted chicken

1 15½-ounce can navy beans or great northern beans, drained and thoroughly rinsed

2 ounces coarsely ground corn tortilla chips

1½ Tbsp. cumin

3 Tbsp. light sour cream

Sprigs of fresh cilantro leaves

Tortilla chips

Yield: 3 servings

Whenever my wife, Christine, and I are in Palo Alto, California, we always try to have breakfast or lunch at the Peninsula Fountain and Grill on Emerson Street. Established in 1923 by the Santana family, this old-fashioned diner always serves very good basic food made from scratch.

One of our favorites for lunch is white bean and turkey chili. Not wanting to wait until our next visit to Palo Alto to enjoy this hearty, healthy dish, I decided to reproduce it at home. Instead of using roasted turkey, which we don't have all that often, I substituted roasted chicken as a way to use leftovers.

There are three key components in the dish that required some experimentation. First is the mixture of roasted chilies. Although the Peninsula Fountain and Grill uses a mix of serrano and jalapeño chilies, the dish is not very spicy. The secret is to remove virtually all of the white pith inside the chilies. Capsaicin, the compound responsible for the pungent flavor of hot chilies, is produced only in the pith and migrates to a small extent to the seeds. Removing both the pith and the seeds eliminates nearly all of the heat. If you like more heat, you can remove less of the pith or add red pepper flakes. I like to grill the chilies and then remove most of the charred skin. You can also char them under the broiler.

The second key component is the less obvious use of coarsely ground corn tortilla chips. They add lots of flavor, as well as thickening the liquid. It took me a few tries to realize this was one of the secret ingredients.

The third component is a flavorful chicken stock. If you use a commercial chicken stock, choose one that is low in salt (some can contain as much as 680 milligrams of sodium per cup). If I am using leftover roasted chicken, then I like to make my own stock with the carcass. That way I can both control the sodium content and ensure fresh flavor. To make 4 cups of stock, use one coarsely chopped carrot, one chopped celery stalk plus some leaves, one chopped medium onion, 1 teaspoon of chopped fresh sage, and 1/2 teaspoon of salt. This is equivalent to 295 milligrams of sodium per cup of stock. Simmer all the ingredients in 4 cups of water, together with the leftover chicken carcass, for about 2 hours; then cool and strain. Add in enough water to make 4 cups.

For this recipe, almost any variety of white beans will work, but we prefer smaller navy beans (the same beans used for Boston baked beans). You can also use great northern white beans. For convenience, I use canned navy beans. Like all canned beans, they tend to be high in salt, which is added both to enhance the flavor and to ensure a uniform tender texture when the beans are cooked in the can. The sodium content of canned beans can be reduced by as much as 40 percent by simply draining and rinsing the beans (Duyff, Mount, and Jones 2011). This simple procedure reduces the sodium content of a 15½-ounce can of beans to 819 milligrams of sodium. Together with the salt in the stock and the tortilla chips, one serving of white bean and roasted chicken chili contains about 570 milligrams of sodium and only 374 calories.

DIRECTIONS:

Slice the peppers in half, and carefully remove the pith and seeds. Char them on the grill (or under the broiler), cool, remove most of the blackened skin, and cut them into ½-inch pieces.

In a 3-quart saucepan, heat the olive oil, peppers, and onion until the onion is translucent (about 5 minutes). Add the garlic, and cook for another 30 seconds (do not overcook the garlic, or it will become bitter). Add the chicken stock, chicken, and beans to the saucepan, followed by the ground tortilla chips. Add the cumin, and mix all the ingredients.

Simmer the mixture for about 20 minutes, until most of the tortilla chips have disintegrated and the liquid has thickened. Taste and season with additional salt and ground black pepper, if needed. If the chili is not spicy enough, add a small amount of red pepper flakes, and simmer for a few more minutes. Top each serving of chili with 1 tablespoon of sour cream, a few sprigs of cilantro and a few tortilla chips. Enjoy!

The good carbohydrates are the nondigestible polysaccharides that function as dietary fiber, providing laxation and the feeling of fullness (satiety) with fewer calories. Whole grains and legumes are the best sources. If you are uncertain about what dietary fiber is, you are certainly not alone. In 2016, after more than 20 years of debate, the FDA finally settled on a definition for dietary fiber and a recommended daily intake of 28 grams per day for adults. The average American consumes only about 16 grams per day. In case you are wondering, the FDA's final definition is as follows: "non-digestible soluble and insoluble carbohydrates (with 3 or more monomeric units), and lignin that are intrinsic and intact in plants; isolated or synthetic non-digestible carbohydrates (with 3 or more monomeric units) determined by FDA to have physiological effects that are beneficial to human health." As a complex polysaccharide, starch is a special case in that it is not a dietary fiber according to the FDA definition because it can be rapidly or slowly digested to maltose and glucose in the small intestine. Thus it elevates the level of glucose in the blood and the secretion of insulin while contributing about 4 calories of energy per gram, which squarely puts these forms of starch in the "bad" carbohydrates camp.

However, a small portion of starch, known as *resistant starch*, is not digested in the small intestine due to its crystalline structure or inaccessibility to digestive enzymes, has a subdued effect on blood glucose and insulin levels, and is very beneficial for the good bacteria living in the large intestine that eagerly metabolize resistant starch to short-chain fatty acids that are used for energy by the cells lining the large intestine. I have conducted research on dietary fiber and resistant starch off and on for about 20 years (Fabbri, Schacht, and Crosby 2016). Resistant starch functions as a prebiotic soluble dietary fiber (although it is an insoluble form of fiber), increases the absorption of calcium into the body, and is believed to reduce the risk of colon cancer and inflammatory bowel diseases of the large intestine. Resistant starch also contributes less than half of the calories of regular digestible starch and is definitely a good carbohydrate that should be included in a healthy diet. Unfortunately, Americans consume only about 3–8 grams of resistant starch per day depending on the diet, compared with a recommended intake of 20 grams per day by some health organizations. The best sources of resistant starch are legumes and whole grains and, to a lesser degree, some extruded cereal products. The levels of resistant starch in raw legumes such as beans, peas, and lentils are actually very high (around 35 percent by weight), but raw legumes must be cooked to be edible, and this destroys all but about 5–6 percent of

(CONTINUED ON PAGE 146)

Winter vegetable! That's what my mother called macaroni when I was growing up in Massachusetts. When vegetables were scarce and prices high during the winter months, there was always macaroni. My favorite was winter vegetable with brown gravy whenever we had roast beef on Sunday afternoons. When we had no brown gravy, then melted butter with salt and pepper was a good stand-in. My mother had lots of other one-dish meals made with versatile macaroni: American chop suey, sausage casserole (made with breakfast sausage links and canned stewed tomatoes), and, of course, macaroni and cheese.

To this day, I am still a great fan of all forms of pasta, not just macaroni. But, unfortunately, pasta has gotten a bad rap for being a high-carb, high-calorie food, and as a result, many home cooks avoid it. In fact, pasta is simple to prepare, versatile, nutritious, and satisfying; it's a food that most adults and children like. One of the most important measures of the impact of carbohydrates on our health is called the *glycemic index* (GI), which is a numerical rating of foods based on their effect on blood sugar levels in our body. Some high-carbohydrate foods, especially those high in starch, cause a sudden surge in the levels of the sugar glucose in the blood; this in turn triggers a rapid release of the hormone insulin, which signals cells to take up the glucose as a source of energy for the cells. If the glucose is not quickly taken up and used for energy, then the insulin shuttles the excess glucose in the blood to fat cells, where it is converted to fat and stored for future use. Thus too many high-carb foods can result in excess fat and weight gain, potentially leading to diabetes and heart disease. Foods that produce a sudden surge in glucose levels have a high GI, while those that release glucose more slowly into the blood have a low GI. It is therefore desirable to eat high-carbohydrate foods that have a low GI.

The GI of a food represents a comparison of the percentage of glucose released into the blood from a specific amount of food to pure glucose, which is assigned a GI of 100. Every food is therefore ranked on a scale of 0 to 100 as determined by analyzing the level of glucose in the blood over a period of 2 hours after consuming a known weight of the food, typically 50 grams (glucose levels are measured over 3 hours for diabetics). Foods with a GI above 55 are considered high GI foods, while those with a GI below 55 are ranked as low GI foods. One serving of a baked potato (4 ounces) has an average GI of 85, 4.3 ounces of French fries have a GI of 75, and 1 cup (6 ounces) of short-grain white rice has a GI of 72, while the same amount of long-grain rice has a GI of only 56. Why is one variety of rice ranked so much lower than another? Recall that starch is made up of two molecules called *amylose* and *amylopectin* (see box 1.2, The Microscopic World of Starch in Food). Amylose readily forms crystalline structures that are more resistant to digestive enzymes than the amorphous forms of amylopectin that are rapidly digested to glucose. The starch in short-grain rice contains high levels of amylopectin and very little amylose compared with long-grain rice. Legumes such as kidney beans, lentils, navy beans, and chickpeas are among the plant foods with the highest levels of amylose, so it is not surprising that the GIs of these foods range from 27 to 38 even though they contain lots of starch. Not all starch is alike, just like not all high-starch foods are fattening.

So what would you guess is the GI of one serving (1 cup, or 6 ounces) of cooked spaghetti? You might be surprised to learn it is only 41, which is clearly in the range of a low GI food. One cup of macaroni (5 ounces) has a

GI of only 45. But it's not because the flour used to make pasta is high in amylose like long-grain rice and beans. Another factor is responsible for the low GI of pasta: the protein content of durum wheat, which is coarsely ground into semolina flour and used to make most pasta. Unlike in the highly refined, finely milled bread flour used to make white bread, the protein and starch granules in semolina flour are largely undisturbed. Both coarse-ground semolina flour and finely ground white bread flour contain about the same amounts of starch (about 73 percent of dry weight), amylose (about 22–28 percent of the total starch), and protein (12–14 percent of dry weight), yet the GI of white bread is 70, while that of spaghetti is only 41. To make dried pasta, coarsely ground semolina flour is mixed with water, made into dough, extruded into spaghetti (or other forms), and dried, leaving the starch granules still intact, surrounded by a protective web of strands of gluten protein (for more on the science of gluten, see box 2.1, Explaining Gluten). When wheat is ground very finely to make refined white bread flour, many of the starch granules are broken apart and the protein matrix fragmented. Unlike in refined white bread, the protective web of protein in pasta prevents the starch granules from bursting and releasing starch molecules when it is cooked. Thus the protein protects the starch in pasta from rapid digestion to glucose, resulting in much slower release of glucose into the blood compared with white bread. But not all pasta is alike because some brands are made with lower-protein flours. If the cooking water becomes very cloudy when the pasta is cooked, this is a sign that unprotected starch granules have burst, releasing molecules into the water and producing pasta that will have a higher GI than better-quality, higher-protein pasta. It's also a good reason to cook pasta only long enough to reach the al dente stage so the starch granules have not all ruptured.

REFERENCES

Brand-Miller, J., T. M. S. Wolever, S. Colagiuri, and K. Foster-Powell. *The Glucose Revolution: The Authoritative Guide to the Glycemic Index*. New York: Marlowe, 1999.

Pagani, M. A., M. Lucisano, and M. Mariotti. "Traditional Italian Products from Wheat and Other Starchy Flours." In *Handbook of Food Products Manufacturing*, ed. Y. H. Hui. Hoboken, NJ: Wiley, 2007.

TABLE 7.1
Resistant Starch in Selected Foods

Food	Mean Grams of Resistant Starch per 100 Grams of Food as Consumed
Granola	0.1
Oatmeal cookies	0.2
All-bran cereal	0.7
Potato, baked	1.0
Whole wheat bread*	1.0
Spaghetti, cooked	1.1
Potatoes, boiled	1.3
Rice, brown, cooked	1.7
Peas, cooked/canned	2.6
Crackers, Ryvita	2.8
Corn flakes	3.2
Rye bread	3.2
Lentils, cooked	3.4
Beans, white, cooked/canned	4.2
Pumpernickel bread	4.5
Oats, rolled, uncooked	11.2

* Containing 51% whole wheat flour.

the resistant starch by weight. The resistant starch that survives is very stable to further cooking; for example, reheated canned beans contain about the same amount of resistant starch as freshly cooked beans. Although 5–6 percent does not sound like a lot, it is among the highest levels in any cooked food and is sufficient to provide ample fuel for the good bacteria in the large intestine that thrive on resistant starch. Levels of resistant starch in a variety of prepared and cooked foods are shown in table 7.1. If you want to increase the resistant starch in your diet, add dry rolled oats to your breakfast cereal. White beans, such as cannellini or lima beans, make a wonderful side dish for roasted chicken when slow cooked in the oven with a whole sliced onion, a little dry mustard and salt, and perhaps a few sprigs of fresh rosemary.

One family of nondigestible polysaccharides that deserves more attention both in cooking and in health is the pectins, which are major structural components of all plant cell walls as well as the glue between cells that binds them together. You are probably familiar with them as the gelling agent that forms jams and jellies. Pectins are less stable in alkaline conditions and more stable under acid conditions. This explains why cooking green beans takes longer in acidic tomatoes than in plain water and why adding a pinch of alkaline baking soda cuts the cooking time in half when boiling green beans and making polenta with coarse cornmeal. The complex polysaccharide structures of pectins can be broken down by an enzyme called *pectin methylesterase* (PME), which is what causes fruits to soften as they ripen. Fruits and vegetables also soften as they cook because heat breaks down their pectins. The technique of precooking can be used to activate PME in fruits and vegetables, which prevents them from becoming mushy when roasted in the oven for too long. PME alters the structure of the pectins, allowing them to form ionic cross-links with calcium ions, thus fortifying the cross-linked pectins against breakdown by heat. The enzyme is most active between 130°F and 140°F (55°C and 60°C) but is deactivated above 160°F (70°C). Fruits and vegetables that contain the highest levels of PME and that are the most benefited by precooking are apples, cherries, beans, cauliflower, tomatoes, beets, carrots, and both white and sweet potatoes. After heating any of these in the oven for about 30 minutes at a temperature between 130°F and 140°F, they can be cooked until tender at higher temperatures (as when roasting carrots) without turning limp or mushy. This technique is often used to advantage when making apple pies, as it prevents the apples from becoming mushy and turning the pie into soup. Apples purchased out of season (after January) that have been in controlled atmosphere storage for many months tend to turn mushy more than fresh apples. If you have ever encountered "hard core" sweet potatoes that refuse to soften in the center when baked in the oven, it is due to the action of PME while the potatoes were stored for long periods in the refrigerator. The PME, together with the calcium ions released, slowly reinforces the cell wall pectins during cold storage, making it impossible to soften the sweet potatoes by cooking.

Like resistant starch, the soluble, nondigestible pectins are eagerly metabolized in the large intestine by billions of healthy bacteria that form the gut microbiome. Mounting evidence is showing that pectins appear to have a number of health benefits. The fact that they are intimately bound up with other polysaccharides in plant cell walls and subsequently broken down into fragments by the gut bacteria makes it very challenging to ferret out their health effects. The gelling behavior of pectins has been shown to lower cholesterol levels in the blood by trapping the cholesterol and eliminating it from the body, and there is some evidence that they may help reduce inflammation and retard the growth of tumor cells. Unfortunately, Americans consume only a little more than half the recommended dietary intake of fruits and vegetables and have done little to change this picture during the last 30 years.

Good Fats and Bad Fats

The average adult consumer appears to have a reasonably good understanding of good carbs and bad carbs, probably because the nutrition community and the media have had a fairly consistent message on carbohydrates and health for the past 20–30 years. The picture is not so obvious for good fats and bad fats, largely because the media, with the food industry's help, have vilified fats and overemphasized low-fat and fat-free foods during the 1990s and well into the following decade. The message they have delivered is that high-fat diets are the cause of overweight, obesity, and cardiovascular disease. Yet nutrition studies have consistently shown it is possible to maintain and even lose weight with diets in which fat constitutes up to 40 percent of the calories. It all comes down to regulating the calories consumed versus the calories burned—and not the source of the calories. Fat has been shown to contain 9 calories per gram (versus 4 calories per gram for carbohydrates and proteins), which probably accounts for the bias that fat makes you fat. Consumers have known about the unhealthy relationship of cardiovascular disease and fats since Dr. Ancel Keys showed a connection between the consumption of saturated fat and heart disease beginning in the 1970s. More recently, the concern has spread to the health effects of cooking oils and the alarm about the smoke point of oils, the temperature at which cooking oils begin to smoke in a hot pan or fryer. The average consumer has been bombarded with different viewpoints on the safety of both fats and oils.

So let's start with the basic facts about the chemistry of fats and oils. In very simple terms, at room temperature, fats are solid and oils are liquid. Sometimes you will see the word *fat* used when referring to both fats and oils, so keep in mind there are solid and liquid forms of fat. Fats and oils are chemically called *triglycerides*, meaning they are composed of three long-chained fatty acids linked to a single molecule of an alcohol named *glycerol*. The fatty acids can be saturated or unsaturated, which refers to the types of carbon-carbon bonds and the number of hydrogen atoms attached to each carbon atom. Saturated fatty acids contain the maximum possible number of hydrogen atoms, so they are therefore "saturated" with hydrogen ($-CH_2-CH_2-$). Unsaturated fatty acids have varying numbers of carbon-carbon double bonds ($-CH=CH-$) that contain fewer hydrogen atoms. As illustrated in box 7.2, Fats and Oils—When Structure Dictates Function, molecules of saturated fatty acids are linear and can pack together like an ordered box of crayons, making them crystalline with a melting point above room temperature. Molecules of unsaturated fatty acids have kinks, which prevent them from forming ordered crystalline structures and therefore enable them to melt below room temperature. Solid saturated fats contain an abundance of saturated fatty acids, while liquid unsaturated oils contain an abundance of unsaturated fatty acids. It stands to reason that an abundance of solid saturated fats can coalesce into crystalline structures within blood vessels (fat in the blood occurs as part of complex particles such as LDL cholestrol.) Oxidation of the crystalline fat lining blood vessels creates insoluble deposits called *plaque* that restrict the flow of blood, including that to the heart and brain. Unsaturated oils have much less tendency to form crystalline structures and are therefore less likely to clog blood vessels even though they are more readily oxidized than saturated fatty acids. Then there are the *trans fats*, which are relatively minor components of natural fats but which are produced in abundance when unsaturated oils are commercially treated with hydrogen gas in the presence of a catalyst, creating unsaturated fatty acids in which the carbon-carbon double bonds assume a new shape (called *trans*, or opposite), as shown in box 7.2. Like saturated fatty acids, trans fats are linear crystalline solids, but they are more readily oxidized than saturated fatty acids. Chemically, trans fats are predisposed to clog blood vessels and form plaque even more readily than saturated fats do.

FATS AND OILS—WHEN STRUCTURE DICTATES FUNCTION

In architecture, we often hear the expression "form follows function," but in chemistry, the opposite is true. The three-dimensional structure of a protein directly determines is function, such as a digestive enzyme or a hormone like insulin. Perhaps one of the best examples of a relatively simple chemical structure that dictates function can be found in fats and oils. Simply put, fats are solids at room temperature, while oils are liquids, and the reason for this is directly related to their chemical structure. The structure of these molecules dictates not only their physical properties such as melting point but also their behavior in the human body, their impact on health, and their reactions to cooking.

Fats and oils belong to a larger family of molecules called *lipids* that also includes steroids like cholesterol, progesterone, and bile acids; the oil-soluble vitamins A, D, E, and K; and the colorful carotenoids such as lycopene and beta-carotene. Chemically, all fats and oils are called *triglycerides* (sometimes called *triacylglycerides*), the chemical name that indicates they have three fatty acids that are attached to a simple alcohol called *glycerol* (the suffix *-ol* identifies it as an alcohol) by linkages called *ester bonds*, which are formed when acids react with alcohols. Although glycerol is present in all triglycerides, the structure of the three fatty acids can differ based on the number and type of carbon-carbon double bonds (C=C) they contain. Fatty acids occurring in plants, animals, and fish can contain anywhere from no double bonds (called saturated fatty acids) to as many as six double bonds (called polyunsaturated fatty acids) contained within long chains ranging from ten to twenty-two carbon atoms (with eighteen carbon atoms being the most common), and each of these acids can be attached at three different positions on glycerol, thus generating fats and oils composed of a very large number of different triglycerides. A vegetable oil like corn oil and a fat like beef tallow are composed of many different triglyceride molecules.

An example of the chemical structure of a single triglyceride molecule is shown in figure 1. The long zig-zag chains in red, green, and blue represent three different fatty acids, while the short black lines connected to the Os in the middle represent a glycerol molecule attached to the three fatty acids by ester bonds (-O-C=O). The double lines within the green and red fatty acids represent carbon-carbon double bonds. Each angle in the zig-zag lines represents the location of a carbon atom. The green and red fatty acids contain a total of eighteen carbon atoms each, while the blue fatty acid contains sixteen carbon atoms (including the C=O and the CH_3 carbon atoms on each end). If you look closely at the carbon-carbon double bonds, you will notice that the carbon atoms on either side of the double bonds reside on the same side and are designated as *cis* double bonds, meaning the carbon atoms attached to the double bond lie on the same side. Virtually all double bonds in naturally occurring fatty acids have the cis configuration, as shown in figure 1. *Trans* double bonds, found in commercially produced, unhealthy trans fats, have the adjacent carbon atoms on the opposite sides of the double bond (the adjacent groups are *transposed*).

MODELS OF A TRIGLYCERIDE MOLECULE

A structural change as simple as the conversion of a cis double bond to a trans double bond can turn a healthy fat into an unhealthy fat. Why is this? Figure 2 shows three molecular models of fatty acids, each containing nine carbon atoms. These are called Dreiding models, which

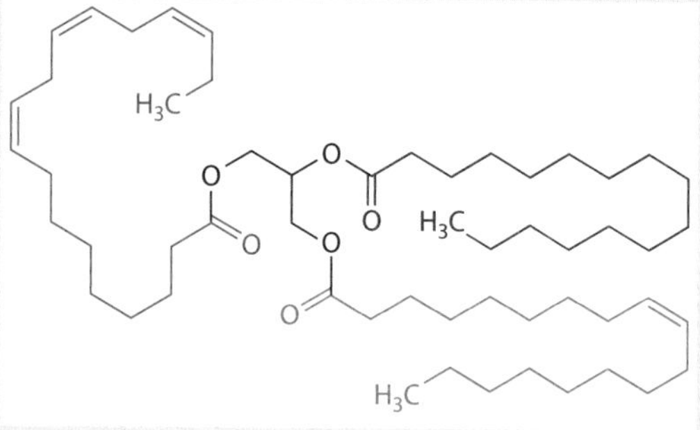

FIGURE 1

The chemical structure of a triglyceride fat molecule.

are manufactured so the bond lengths and bond angles are very precise in terms of actual atomic dimensions.

On the right end of each model, you will see the acid group, called a *carboxyl group*, which contains two red oxygen atoms. The molecule at the top is a saturated fatty acid in which the black carbon atoms with attached hydrogen atoms adopt a linear zig-zag shape. The molecule in the middle is a trans-fatty acid in which the lone carbon-carbon double bond starts at the third carbon atom from the carboxyl group. Notice that the two hydrogen atoms attached to the double bond are trans (opposite) to each other. But, most important, the linear zig-zag shape of the trans molecule is almost identical to the zig-zag shape of the saturated fatty acid just above. Saturated fats are known to build up inside blood vessels and become oxidized, forming hard deposits of plaque that block blood vessels and cause cardiovascular disease. The nearly identical shape of the trans fats causes them to do the same thing, but because unsaturated trans fats are more easily oxidized, they form artery-clogging plaque even more readily than saturated fats do.

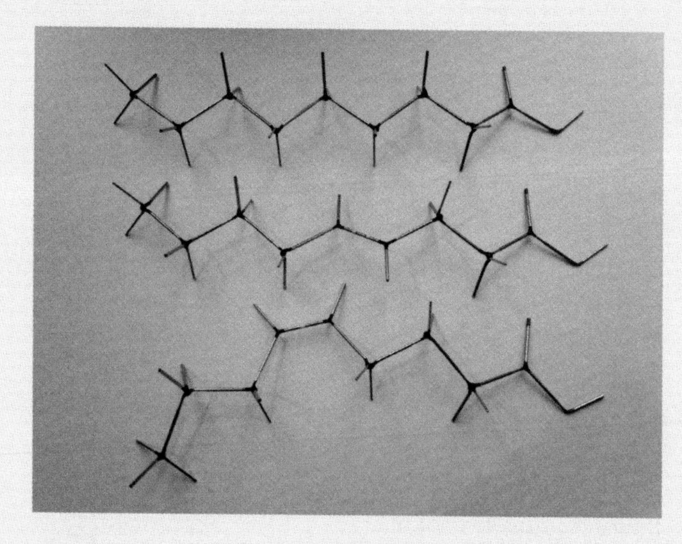

FIGURE 2

A comparison of precise three-dimensional Dreiding molecular models of a saturated fatty acid (top), a trans fatty acid (middle), and a cis fatty acid (bottom). The saturated and trans fatty acids possess similar three-dimensional structures that differ from the three-dimensional structure of the cis fatty acid. Photograph by the author.

The bottom molecule in figure 2 contains a cis double bond that starts at the fourth carbon atom from the carboxyl group. Notice that the two hydrogen atoms attached to the carbon atoms of the double bond lie on the same side of the double bond. Notice also that the shape of the molecule is no longer linear because the cis double bond causes a bend in the zig-zag shape of the molecule. Saturated fats and trans fats have a linear shape that allows the fatty acids to line up with each other like a box of crayons and form organized crystalline structures that melt above room temperature, so they are solids like beef tallow and lard; naturally occurring cis fatty acids such as those in unsaturated vegetable oils cannot easily form crystalline structures, so their melting points are below room temperature. Changing a double bond from a trans configuration to a cis configuration changes a solid fat into a liquid oil. Unsaturated fatty acids are less able to form solid plaque in arteries than saturated and trans fats are. The classic example in which fat dictates function is cocoa butter, the highly saturated fat in chocolate: it can exist in six different crystal structures, but only one is responsible for the smooth sheen and snap of good-quality chocolate, as well as its very important property of melting in your mouth and not in your hand.

Finally, you have probably heard of omega-3 and omega-6 fatty acids and the beneficial role of omega-3 fatty acids in health. The numbers 3 and 6 refer to the location of the last double bond in the unsaturated fatty acid. In the triglyceride molecule in figure 1, the last double bond in the red omega-3 fatty acid (linolenic acid) starts at the third carbon atom from the end of the zigzag chain of carbon atoms (*omega* is the last letter in the Greek alphabet, so it designates the end of the chain), while the lone double bond in the green fatty acid is nine carbon atoms from the end of the chain, meaning it is an omega-9 fatty acid, oleic acid, which is found in abundance in olive oil. The last double bond in an omega-6 fatty acid such as linoleic acid (not pictured) starts at the sixth carbon atom from the end of the chain. The best sources of heart-healthy omega-3 fatty acids are fatty fish (especially salmon, tuna, and sardines), walnuts, almonds, and flaxseed. Omega-6 fatty acids are obtained in abundance from vegetable oils.

TABLE 7.2
Level of Saturation of Common Fats and Oils

Fat or Oil Source	Iodine Value*	% Saturated	% Unsaturated
Butter	33	65	35
Beef (tallow)	47	48	51
Palm oil	52	49	51
Pork (lard)	56	42	58
Chicken	77	33	67
Olive oil	82	12	88
Peanut oil	92	18	81
Canola oil	107	7	92
Corn oil	123	14	86
Soybean oil	131	15	85
Herring oil	162	21	79

* Numbers are average iodine values.

As I will discuss a little later in this chapter, there are many reasons to choose healthier unsaturated oils from plants and fish over saturated fats, especially fat from red meat. Table 7.2 shows the level of saturation of a variety of fats and oils, using the old but reliable *iodine value test* to determine the number of carbon-carbon double bonds contained in the fatty acids (Stauffer 1996). More-saturated fats with the least number of carbon-carbon double bonds have lower iodine values and appear at the top of the table; more-unsaturated oils with more double bonds are found at the bottom. It is clear from these numbers that animal fat is more saturated than vegetable oils and that chicken fat is more unsaturated than beef or pork fat. If you look carefully at the numbers in the table, you will notice there is not a perfect agreement between the iodine value and the percentage of unsaturated fat or oil. This is because the iodine value measures the critical number of double bonds in the fatty acids and not simply the percentage of unsaturated fatty acids by weight in the fat or oil. Olive oil, for example, is about 79 percent monounsaturated oleic acid (one HC=CH double bond), while a number of the unsaturated fatty acids in herring oil have three to six double bonds each. Thus the iodine value of herring oil is double that of olive oil even though the percentage of unsaturated fat in herring oil is lower than in olive oil.

When it comes to good fats and bad fats, the clear villain in the crowd is trans fat. Because its link to heart disease has been so clearly proved, the FDA now requires that the content of trans fat in processed food be listed on the Nutrition Facts label. Although a very small amount (1–2 percent) of trans far occurs in both animal fat and refined vegetable oils, the vast majority is commercially produced by the partial hydrogenation of vegetables oils such as soybean oil. If you see the words *partially hydrogenated vegetable oil* in the list of ingredients on the label of a processed food product, you know it contains trans fat. In 2015, the FDA removed partially hydrogenated vegetable oil from its list of "generally recognized as safe" ingredients approved for use in food. Trans fats are generally solid at room temperature and more resistant to oxidation than unsaturated vegetable oils, so they proved to be useful for deep-fat frying and for the production of solid spreads like margarine to replace solid shortening in baking (and eventually replaced much of the solid saturated fat like tallow and lard used in these applications). In the 1990s, evidence from the Harvard T. H. Chan School of Public Health and many other research institutions began to show the connection between the consumption of trans fat and heart disease. Trans fat increases the level of bad LDL cholesterol and lowers the level of good HDL cholesterol in the blood, promoting the formation of blood clots that can trigger heart attacks and strokes. The ensuing campaign to remove trans fat from processed food and frying fat has been quite successful, with most of it now replaced by blends of solid fats with liquid oils. The World Health Organization has requested that all governments around the world eliminate the use of trans fats by 2023.

Saturated fats are also considered bad fat because they raise the level of bad LDL cholesterol and are therefore associated with an increased risk of heart disease. They are not as bad as trans fat because they do not lower the level of good HDL cholesterol, but they should still be limited in the diet.

The mono- and polyunsaturated oils from plants and fish are considered the good fats. As mentioned earlier in this book, humans require the presence of two chemically related polyunsaturated fatty acids in our diet—linoleic acid, an omega-6 fatty acid, and linolenic acid, an omega-3 fatty acid (the numbers 3 and 6 refer to the location of the last carbon-carbon double bond in the fatty acid). These fatty acids are essential, as they are responsible for producing many important biologically active molecules such as the *eicosinoids*, which are involved in regulating processes such as smooth muscle contraction and relaxation, blood clotting and blood thinning, the production and reduction of inflammation, and much more. Strangely, humans never evolved the enzymes to produce these acids, so they are essential in our diet. They are found in some vegetable oils such as soybean and canola oils, nuts such as walnuts, and green leafy vegetables. Linolenic acid is also very important because the body converts it in very small amounts to two very important longer-chain omega-3 fatty acids you may have heard of, called *eicosapentaenoic acid* (EPA) and *docosahexaenoic acid* (DHA). Both are essential components of cell membranes throughout the body, serving as gatekeepers for what is transported into and out of cells, and DHA is the most abundant fatty acid in the brain. These two important acids are fairly abundant in fatty fish such as salmon, sardines, and mackerel. Don't be put off by farmed salmon, which may contain twice the amount of EPA and DHA as wild-caught salmon due to the diet fed to farmed salmon.

Now let's turn our attention to what happens with liquid oils and solid fats when we cook with them. During cooking, oils and fats can undergo at least three important chemical changes: (1) oxidation, (2) polymerization, and (3) breakdown of the triglyceride molecules to free fatty acids and glycerol. Oxidation of the fatty acids in oils and fats during exposure to high heat in the presence of air produces the delicious taste and aroma of fried food by forming very small amounts of dozens of new compounds called *aldehydes*, *ketones*, and *alcohols*. The most abundant compound formed is 2,4-decadienal, an aldehyde that is responsible for most of the enticing flavor of fried food. Polymerization of cooking oils occurs when they are exposed to high heat and air during sautéing and deep-fat frying. You may have noticed the golden yellow sticky substance that forms a ring around the inner rim of a sauté pan when you are cooking meat or vegetables; it is very difficult to remove without scrubbing with soap and hot water, often aided by an abrasive like baking soda. This is oxidized, polymerized vegetable oil, and it may account for as much as 25 percent of used cooking or frying oil depending on how long the oil was heated. And, finally, high heat in the presence of water or particulate matter from food causes the triglycerides in vegetable oils and fat to break apart into free fatty acids and glycerol even more readily. All of these products, including some small amounts of colored pigments, form and build up during the prolonged heating of oil, especially in oil used for extended periods of time for deep-fat frying. There is nothing more disappointing than being served fried foods with off-flavors and -colors because the restaurant has not changed the cooking oil frequently enough. It is a sure sign of poor quality.

If cooking oil is heated to too high a temperature, the free glycerol will lose two molecules of water and form acrolein, an acrid-smelling, toxic substance that is responsible for the blue haze, or appearance of smoke, above oil heated to its smoke point. The "smoke" from pure cooking oil is actually polymerized acrolein, plus perhaps some water vapor if there is any moist food in the pan. Pure acrolein has a very low boiling point, around 125°F (52°C), so it does not remain in the hot oil. The U.S. Occupational Safety and Health Administration has set a limit of 0.1 parts per million of acrolein vapor in the air inside a commercial kitchen. This is roughly equivalent to 1 second in 2 weeks, which should give you some idea of the toxicity of acrolein.

In my opinion, there is no need to heat cooking oils to their smoke point (usually 400°F/204°C or higher). After all, as soon as you put a piece of moist meat in the pan, the sudden evaporation of water from the surface of the meat will immediately lower the temperature of the oil way below its smoke point. To sear meat, it is much wiser to thoroughly dry the surface of the meat (by using paper towels or putting it in a very dry freezer) and then place the meat in oil hot enough to shimmer but not smoke (about 390°F/199°C). You will get better results with much less splattering of hot oil all over the stove. The best indicator that oil is ready for sautéing is the "shimmer" you see on the surface, the result of convection currents in the oil caused by the heat. The presence of food and water greatly reduces the smoke point of cooking oil, so using the oil for a prolonged period or cooking large amounts of food in batches without changing the oil eventually causes the oil to break down and smoke at a much lower temperature. For example, as fresh soybean oil breaks down to free fatty acids and glycerol during heating, the formation of as little as 1 percent of free fatty acids (by weight of the oil) will lower the smoke point from 414°F (212°C) to 314°F (152°C). Some lower grades of extra virgin olive oil (EVOO) can be as much as 0.5 percent free fatty acids, which causes them to smoke at lower temperatures than highly refined vegetable oils. Very high quality EVOO with very little free fatty acid has a smoke point (405°F/207°C) higher than that of even highly refined canola oil (400°F/204°C). And refined grades of olive oil can have smoke points (468°F/242°C) higher than even those of refined corn, palm, and peanut oils (450°F/232°C). So don't be fooled by people who say you shouldn't cook with olive oil because it has such a low smoke point; it all depends on the grade of the olive oil. One last point about the smoke points of oils: they are not easy to determine accurately unless using the proper equipment, as when shining an intense light across the surface of the oil to make it easier to detect the particles of "smoke" at a specific temperature. So be very cautious when making decisions based on reported smoke points that vary from source to source.

COOKING WITH OLIVE OIL

Without hesitation, I can state that olive oil is my favorite oil for cooking, especially for panfrying (sautéing) almost anything. So why don't more people use olive oil for cooking in view of its "good for you" reputation? Let's find out what the experts think.

Several years ago I asked one of the senior teaching chefs at the prestigious Culinary Institute of American why he didn't recommend olive oil for cooking. He emphatically answered, "Because olive oil imparts too much flavor to food!" "Are you serious? That's a complete myth," I shouted. When I was the science editor for America's Test Kitchen, one of the cooks had done a simple test showing that, when a good grade of extra virgin olive oil (EVOO) was heated for 10 minutes at 350°F (177°C) and then cooled, its taste was indistinguishable from that of refined soybean oil subjected to the same treatment (as determined by a taste panel of ten judges). All of the volatile, green, grassy aromas of fresh olive oil had evaporated, leaving behind oil that was as bland tasting as a highly refined vegetable oil. And that should not be surprising, as two of the dominant aroma compounds, hexanal and Z-3-hexenal, boil at 266°F (130°C) and 259°F (125°C), respectively. Panfrying food with EVOO does not produce food tasting like fruity, grassy, green olive oil. The cooks then compared their favorite brand of EVOO with a commercial blend of refined vegetable oils. The two oils were indistinguishable when cooked with a tomato sauce and oven-roasted potatoes. Although one may argue this proves a good grade of EVOO may not be worth its high price for cooking, I take comfort in knowing that its high levels of phenolic antioxidants, like hydroxytyrosol and tyrosol, and the unique anti-inflammatory agent oleocanthal ensure that more of them survive the cooking process. So I always reach for EVOO for my cooking (although I buy a lower-priced, good-quality EVOO for everyday cooking).

Due to their chemical structure, all vegetable oils are subject to at least three changes during cooking: (1) oxidation, (2) polymerization, and (3) hydrolysis by the water in food. The extent of these reactions is dependent on the composition of the fatty acids in the oils. The very high level of monounsaturated oleic acid in olive oil reduces its susceptibility to oxidation compared with vegetable oils containing more polyunsaturated fatty acids such as soybean, sunflower, corn, and canola oils. The dominant oxidation product of linoleic acid in these oils is 2,4-decadienal, which is produced at a level 3.4 times higher in canola oil and almost 4.5 times higher in soybean oil compared with regular olive oil when the oils are heated at 356°F (180°C) for 15 hours. This is important because of the health concerns regarding the oxidation of cooking oils.

Virgin olive oil is rich in phenolic antioxidants such as hydroxytyrosol and tyrosol and their derivatives of elenolic acid, which help protect the oil from oxidation during heating. Approximately 40–50 percent of the hydroxytyrosol phenols in virgin olive oil are lost after frying one batch of potatoes for 10 minutes at 356°F (180°C), and nearly 90 percent are lost after frying six batches (60 minutes total frying time). Tyrosol and its derivatives are much more stable, losing only 20 percent after frying twelve batches. These results demonstrate that the phenolic antioxidants in virgin olive oil are doing their job protecting the fatty acids from oxidation compared with other vegetable oils. Olive oil is unique in that it is the only vegetable oil that contains the potent anti-inflammatory agent *oleocanthal*. Despite its simple chemical structure, oleocanthal is quite stable to cooking; when

EVOO containing 53.9 milligrams of oleocanthal per kilogram of oil was heated at 464°F (240°C) for 90 minutes, instrumental analysis showed it lost only 16 percent of its oleocanthal. However, the same study showed that up to 31 percent of the biological activity of oleocanthal was lost based on a taste bioassay.

Another reason home cooks won't use olive oil for cooking is that they are concerned about the low smoke point of the oil. They may not be aware that the smoke point of olive oil varies with the degree of refinement of the oil. The smoke point of an oil is reached when the triglyceride molecules break apart to glycerol and free fatty acids and the glycerol is rapidly dehydrated to the toxic aldehyde acrolein, which is responsible for the blue haze seen above "smoking" oil. Unfiltered EVOO can have a smoke point as low as 375°F (191°C). The smoke point of high-quality cold-pressed EVOO containing very little free fatty acid is about 405°F (207°C), while that of refined olive oil is about 468°F (242°C). Except for a few cases, it is not necessary to heat cooking oil to its smoke point. I recommend heating any oil no higher than 374°F (190°C).

While playing around in the kitchen, Christopher Kimball, founder of the new Milk Street Kitchen, noticed that eggs could be scrambled to a light, fluffy consistency with hot EVOO. Eggs leavened by steam turn out to be light and fluffy when the proteins unfold and cross-link at lower temperatures. As the new science editor for Milk Street, I did a little literature searching and found that olive oil contains some unique phospholipid surfactants that presumably help the egg proteins to denature and lightly cross-link at lower temperatures and to form light fluffy curds rather than the overly cross-linked tough scrambled eggs that result from overcooking. This interesting property may lead to new applications for olive oil in making emulsions, sauces, and baked goods.

REFERENCES

America's Test Kitchen. "Does It Pay to Cook with Extra-Virgin Olive Oil?" *Cook's Illustrated* (January & February 2014): 16.

Cicerale, X., A. Conlan, N. W. Barnett, A. J. Sinclair, and R. S. Keast. "Influence of Heat on Biological Activity and Concentration of Oleocanthal—A Natural Anti-inflammatory Agent in Virgin Olive Oil." *Journal of Agricultural and Food Chemistry* 57 (2009): 1326–1330.

Crosby, G. "Do Cooking Oils Present a Health Risk?" *Food Technology* 72, no. 5 (2018): 50–57.

Gómez-Alonso, S., G. Fregapane, M. D. Salvador, and M. H. Gordon. "Changes in Phenolic Composition and Antioxidant Activity of Virgin Olive Oil During Frying." *Journal of Agricultural and Food Chemistry* 51 (2003): 667–672.

Katragadda, H. R., A. Fullana, S. Sidhu, and A. A. Carbonell-Barrachina. "Emissions of Volatile Aldehydes from Heated Cooking Oils." *Food Chemistry* 120 (2010): 59–65.

In recent years, controversy has been building around the health concerns related to cooking oils. The opinion of those who study the chemistry of cooking oils is mixed, with some believing the oxidation products formed in cooking oils present significant health risks and others believing humans have sufficient detoxifying mechanisms to protect against the harmful effects of oxidized oil. In my opinion, there is very little health risk in cooking with vegetable oils in the home because the oils are usually used only once and not heated for very long. It is also possible to control the temperature of the oil and not let it reach the smoke point, where harmful products like acrolein can form. The optimum temperature for cooking oil is about 356°F (180°C) and should not exceed 374°F (190°C). The risk comes in consuming food prepared outside of the home that has been deep-fat fried in cooking oil that has not been changed for several days or more, allowing the buildup of oxidation products and colored pigments. The color and smell of the oil are often telltale signs of oxidized oil. Typical deep-fat batch fryers are operated at least 8 hours per day and the oil filtered every day to remove particulate matter from food, which lowers the smoke point and makes the oil more susceptible to oxidation. But many restaurants fail to replace the used oil with fresh oil more than two or three times per week, and often they may simply add a portion of fresh oil. A recent study of frying oils used in independent (nonfranchise) restaurants found that 35 percent of the restaurants exceeded in-use quality standards. Generally, oil quality cannot be maintained in batch fryers with a turnover rate greater than or equal to 20 hours of operation. Research conducted by the Department of Nutrition at the Harvard T. H. Chan School of Public Health strongly suggests that consuming deep-fat fried foods four or more times per week (mostly outside of the home) significantly increases the risk of developing chronic diseases such as type 2 diabetes, heart failure, obesity, hypertension, and coronary artery disease. Over the past 40 years, consumption of food away from home has increased 42 percent. So don't stop using cooking oils at home, but do limit the amount of fried food consumed outside of the home where you have no control over the quality of the cooking oil.

I advise using a variety of oils at home, including olive, canola, corn, peanut, and soybean oils, so you obtain a mix of polyunsaturated omega-3 and omega-6 fatty acids along with monounsaturated fatty acids like oleic acid. For my one go-to cooking oil, I use EVOO most of the time, although for stir frying I like to use peanut oil. Some cooks argue that they do not like the strong grassy flavor of olive oil. But did you know that, after heating olive oil at 350°F (177°C) for only 10 minutes, all of the volatile grassy notes are driven off and the oil tastes the same as heated refined soybean oil! Olive oil, especially EVOO, is oxidized much more slowly than polyunsaturated oils like canola and soybean oils. Using 2,4-decadienal as an indicator of oxidation, polyunsaturated canola oil produces about 3.4 times more of this aldehyde than EVOO, while soybean oil produces about 4.5 times more under the same cooking conditions. The monounsaturated oleic acid in olive oil is oxidized about ten to twenty-five times more slowly than the polyunsaturated linoleic and linolenic acids in canola and soybean oils, plus olive oil contains significant levels of polyphenols, which act as very effective antioxidants that protect the fatty acids from oxidation. In a well-controlled study using EVOO to fry French fries at 356°F (180°C), there was a threefold decrease in the antioxidant activity of the oil after six consecutive fryings of 10 minutes each (1 hour total frying time) using the same batch of oil (the oil was cooled between each frying), indicating the antioxidants were doing their job protecting the fatty acids against oxidation. If you are concerned with the potential health effects of oxidized cooking oils, then I recommend you use EVOO like I do. The famous PREDIMED diet study conducted with 7,447 women and men in Spain (ages fifty-five to eighty), all at high risk of developing cardiovascular disease due to overweight, high blood pressure, and cholesterol levels, showed that a Mediterranean-style diet containing EVOO (1 liter per week per household) reduced the actual incidents of cardiovascular events (heart attack, stroke, and death) by 30 percent compared with the lower-fat control diet over the course of the 4.8 years of the study. Consumption of mixed nuts (30 grams/day/person) showed similar beneficial results.

Are There Good Proteins and Bad Proteins?

Humans rapidly absorb a simple sugar like glucose and store it in the body as a complex molecule called *glycogen*, which can be converted back to glucose as needed for energy. Humans also absorb fatty acids from the digestion of fats and oils and store them as fat in adipose tissue unless burned for energy. But proteins are not absorbed into the body. Instead, they are broken down by digestion into individual amino acids that are absorbed into the body and used to build the new proteins required to produce essential components such as muscle and connective tissue, the hemoglobin in blood, a multitude of enzymes, and hormones like insulin. In all, there are about ten thousand different proteins created in the body from the amino acids released by digestion of the proteins in our diet.

The proteins in our diet are the source of the twenty different amino acids needed to build the proteins programmed for synthesis by the DNA in our chromosomes when needed by our body to maintain good health. The body is capable of synthesizing eleven of the twenty amino acids as needed but cannot make the other nine, which means they must come from the proteins in our diet. Sources of dietary proteins that contain sufficient quantities of all twenty amino acids, including the essential ones we cannot make, are called *complete proteins*, while those proteins that do not contain enough of the nine essential amino acids to meet our needs are referred to as *incomplete proteins*. It's not the case that these proteins do not contain any of the essential amino acids; it's just that they don't contain them in the amounts our body requires. Sources of complete proteins include eggs, meat, poultry, fish, and dairy products, while many of the proteins in plants, including legumes and cereal grains, are incomplete. Fortunately, the essential amino acids that are deficient in some plant products like rice and corn are sufficient in others like beans and bread, which explains why certain combinations of foods such as beans and rice are a healthy source of all the essential amino acids. In one sense, we might say that complete proteins are the good proteins, but it's not fair to say that the incomplete proteins are the bad proteins because most people, especially those in the developed world, consume about twice as much protein as they need to supply all of the amino acids to maintain a healthy body.

If we were to single out a specific protein as a bad protein—like trans fat is the villain of all fats and added sugars are the bad sugars—then *gliadin*, the cause of celiac disease, would be the clear choice for the bad protein. Gliadin occurs primarily in wheat, as well as rye and barley, and is one of the structural components of gluten, which is formed when water is added to wheat flour and kneaded into dough. As a protein, gluten does not occur in wheat; it is formed only when gliadin chemically combines with another protein in wheat called *glutenin* during the preparation of bread dough. That's why gliadin and glutenin are often referred to as the gluten-forming proteins in wheat. Celiac disease is an autoimmune response to specific sequences of amino acids in gliadin that destroy the villi that line the gastrointestinal tract, resulting in decreased absorption of nutrients accompanied by gas, bloating, stomach pains, diarrhea, and weight loss. It is a terrible disease that is triggered by exposure to even tiny amounts (about 50 milligrams) of gliadin. About one out of every one hundred Americans tests positive for celiac disease. The formation of gluten in wheat dough does not destroy the sequence of amino acids in gliadin that causes celiac disease , so those with celiac disease are sensitive to all sources of gluten in food. Many more suspect they are "gluten sensitive," but this is very difficult to determine without proper testing. The only solution is to go on a gluten-free diet, which is very difficult to accomplish, given the enormous number of food products that contain gluten. Fortunately, many gluten-free products are now available in the supermarket, and many cookbooks provide recipes for preparing gluten-free foods at home. Fairly recently, Spanish scientists developed a genetically modified form of wheat with up to 97 percent less gliadin than regular wheat.

Most proteins occur in their native biologically active state in two basic shapes, globular and fibrous. The long chains of amino acids in globular proteins, such as insulin, hemoglobin, eggs, and a plethora of enzymes, are twisted into coils resembling a somewhat spherical shape, while the amino acids in fibrous proteins, including actin and myosin (muscle proteins) and collagen (the dominant protein in connective tissue), are arrayed in a somewhat linear shape. When heat is applied to proteins, they undergo a process known as *denaturation*, causing the globular proteins to unfold and the fibrous proteins to contract or shrink. This is what happens to the globular proteins in an egg and the fibrous proteins in muscle tissue in meat and fish when they are cooked. The numerous proteins in egg whites and yolks unfold and form chemical cross-links with each other, creating an infinite network of united proteins that form the rubbery solid structure of a cooked egg. None of the water in a raw egg is lost when it is cooked; rather, it becomes trapped inside the infinite network of proteins. When meat is cooked, the fibrous muscle and connective proteins shrink in both diameter and length, squeezing out some of the water trapped within the network of proteins and causing overly cooked meat to become dry. The loss of water from meat can be reduced by brining the meat in a dilute solution of salt (sodium chloride) before cooking. The salt slowly diffuses into the meat and dissolves some of the muscle proteins. During cooking, the dissolved proteins form a gel that holds the water in the meat, much like gelatin released from collagen in connective tissue.

The process of denaturation occurs at a very specific temperature for each protein. For example, egg whites (10 percent protein and 90 percent water) from hen eggs begin to coagulate and thicken around 145°F (63°C) and are fully firm and rubbery at 176°F (80°C), while the yolks (16 percent protein, 35 percent fat, and 54 percent water) begin to thicken at about 150°F (65°C) and become fully set at 172°F (78°C). Because egg white proteins set over a wider temperature range than egg yolk proteins, cooking whole eggs with the desired texture can be a challenge. The muscle and collagen proteins in meat have a similar specific set of temperatures at which they begin to shrink. Surprisingly, the muscle proteins in beef start to shrink at temperatures as low as 104°F (40°C), with maximum shrinkage and water loss occurring at 140°F (60°C) and above. The purpose of citing all these temperatures is to illustrate how important it is to carefully control the temperature of the food within a specific range, as well as the cooking time, when cooking meat, poultry, fish, and eggs.

Cooks commonly brine and marinate meat, poultry, and fish to increase tenderness, retain moisture, and enhance taste. Brining works by allowing salt to diffuse into the food, where the increased levels of sodium and chloride ions dissolve some of the muscle proteins, enhancing tenderness and forming gels during cooking that bind water and aid in moisture retention. The heightened level of salt also enhances the taste of the food. Marinating food is thought to perform similar functions, but research has shown that the acids in marinades, such as vinegar and lemon juice, penetrate very poorly, producing effects mostly near the surface of the food. Only marinades that contain high levels of salt, such as that found in soy sauce, will penetrate beyond the surface.

The diffusion of a substance is defined as the quantity of molecules or ions that in a unit of time (usually a second) pass through a unit of area (usually a square centimeter) of a plane perpendicular to the direction of movement. The diffusion of molecules or ions within another substance, such as a gas, liquid, or solid, occurs by the random motion of the molecules or ions, known as *Brownian motion*. Brownian motion is caused by the collision of particles, or molecules, with the atoms or molecules that make up the gas, liquid, or solid. Calculating how far salt or acid diffuses into meat, poultry, or fish after a certain amount of time is not too different from calculating how far a car can go in 3 hours if it travels at an average speed of 50 miles per hour. The equation for this calculation is distance = rate × time. If the car is traveling at an average rate of 50 miles per hour for a time of 3 hours, then the car will travel $50 \times 3 = 150$ miles.

Calculating how far a car will travel is fairly straightforward because the car travels on defined roads in specific directions. It can't go just anywhere unless it's an off-road vehicle. Molecules or ions that diffuse into meat or fish are more like off-road vehicles moving in all directions at random, similar to Brownian motion. So we can't use a formula as simple as distance = rate × time to calculate how far a large number of molecules or ions will randomly move in a certain amount of time. But one characteristic of the diffusion of molecules or ions helps to simplify the equation; for a very large number of molecules or ions, diffusion always occurs from a high concentration of molecules or ions to a lower concentration. Placing meat or fish in a concentrated brine or marinade will cause the molecules or ions to diffuse into the meat or fish, where the concentration of molecules or ions is much lower.

To determine how far salt or acid will diffuse into meat or fish within a certain amount of time, we need to measure the *diffusion coefficient* of the substance in the food. The diffusion coefficient (D) is a measure of the rate of movement of that substance (salt or acid). It is similar to the miles per hour of the car, which we simply read on the car's speedometer. While the diffusion coefficient is a number that must be measured just like the speed of the car, it is expressed in units of square centimeters (cm^2) per second instead of miles per hour. In other words, D is a measure of how fast molecules or ions (in the case, of salt or acid) move through a unit of area (square centimeters) in a unit of time (seconds).

Diffusion coefficients must be measured in the laboratory. For substances such as salt and acid, there are a number of laboratory methods that can do this. For each specific food, the diffusion coefficient must be measured under well-defined conditions of concentration, temperature, and time, as these variables will affect the result. Let's take the example of beef. A recent paper published in the journal *Meat Science* (Lebert and Daudin 2014) reported that the *mean* diffusion coefficient (*D*) for a salt solution diffusing into beef measured at various concentrations (1.5–10 percent) at various times (2, 4, and 6 days) was 5.1×10^{-6} cm^2/second at 50°F (10°C)—this is the same as 0.0000051 cm^2/second. It shows the rate at which sodium and chloride ions move through a very small area in a very brief amount of time, similar to how cars move through a tollbooth. The same paper stated that the *mean* diffusion coefficient (D) for protons (H^+, acid) formed by the ionization of acetic acid was 3.5×10^{-7} cm^2/second at 50°F (10°C) for various concentrations and times. Note that this is more than ten times lower than the rate at which the sodium and chloride ions move (which both move at the same rate). The diffusion coefficient for protons in beef is about sixty times lower than that for protons in pure water because the protons bind to proteins, especially the proteins in connective tissue, and this impedes their motion.

A much earlier study published in the *Journal of Food Science* (Rodger et al. 1984) found that the diffusion coefficients of salt and protons into herring were 2.3×10^{-6} cm^2/second and 4.5×10^{-6} cm^2/second, respectively. In this case, the rate of diffusion of acid is about twice that of salt. This is because fish contain very little connective tissue to impede the movement of protons, which normally diffuse faster in pure water than salt does.

Calculating how far salt and acid (as protons, H^+) will diffuse into meat or fish after a certain amount of time requires a more complicated equation than distance = rate × time. The equation is based on the concept of a *random walk*, a term first proposed by Karl Pearson in 1905. Pearson wanted to know how an infestation of mosquitoes spreads in a forest, so he invoked the concept of a random walk to describe the motion of the mosquitoes. That very same year Albert Einstein published a famous paper on Brownian motion in which he calculated the complicated path of a dust particle in air based on the concept of a random walk driven by collisions of the dust particle with gas molecules in the air. This is very similar to the diffusion of salt or acid in water. For his solution, Einstein focused on the calculation of the diffusion coefficient of a dust particle moving in air by Brownian motion. Thanks to Einstein, we can use a simplified equation for a random walk and the diffusion coefficient for salt or acid to calculate how far salt or acid will diffuse into beef after a certain amount of time.

The simple equation we use to calculate the random walk of salt or acid into meat or fish is $L = \sqrt{4 \times D \times t}$, which means the square root ($\sqrt{\ }$) of 4 times the diffusion coefficient (*D*) times the time (*t*, in seconds). L = the distance the salt or acid moves into meat or fish in a given time *t*. Why do we calculate the square root of the product of $4 \times D \times t$? Because the diffusion coefficient is expressed as cm^2 over time, and we want to calculate the distance in centimeters, not square centimeters (an area). The square root of cm^2 equals centimeters. (For example, $3 \times 3 = 9$, while the square root of $9 = 3$.) So let's calculate some examples.

1. How far will salt diffuse into beef in 3 hours?

First, 3 hours = 10,800 seconds (3 hours × 60 minutes × 60 seconds) = 10.8×10^3

Answer: $L = \sqrt{(4 \times 5.1 \times 10^{-6} \text{cm}^2/\text{sec} \times 10.8 \times 10^3)} = \sqrt{0.22} = 0.5$ cm

2. How far will acid diffuse into beef in 3 hours?

Answer: $L = \sqrt{(4 \times 3.5 \times 10^{-7} \text{cm}^2/\text{sec} \times 10.8 \times 10^3)} = \sqrt{0.015} = 0.1$ cm

3. How far will acid diffuse into fish in 3 hours?

Answer: $L = \sqrt{(4 \times 4.5 \times 10^{-6} \text{cm}^2 / \text{sec} \times 10.8 \times 10^3)} = \sqrt{0.19} = 0.4$ cm

Does this agree with what we observe? Let's consider brining meat in a 5 percent salt solution versus marinating meat in vinegar that is 5 percent acetic acid. Both contain equal weights of salt and acetic acid. But in solution, salt is completely ionized to sodium and chloride ions, so a 5 percent solution of salt contains 50 grams of sodium and chloride ions per liter. Acetic acid is a weak acid, meaning only a tiny fraction is ionized to hydrogen ions [H^+]. The strength of acids is measured by the concentration of hydrogen ions in solution, which is what affects the proteins in meat and fish. The pH of vinegar = 3, which equates to only about 0.001 grams of hydrogen ions per liter. So a 5 percent solution of acetic acid (vinegar) contains only 0.001 grams of hydrogen ions liter. This is far less than the 50 grams of sodium and chloride ions per liter.

Because of the concentration differences between the ions of sodium and chloride and the protons in a weakly ionized acid, salt should diffuse into food much faster than acid. And, in fact, almost no protons diffuse into meat compared with sodium and chloride ions. Does it now make more sense that so little acid, in the form of hydrogen ions, penetrates meat or fish compared with salt, in the form of sodium and chloride ions? It's not only because hydrogen ions (protons) diffuse into meat much more slowly than sodium and chloride ions but also because the concentration of hydrogen ions from vinegar is exceedingly small. (Lemon juice has a pH = 2, which is equivalent to only 0.01 grams of protons per liter. This is ten times higher than the concentration of vinegar but still a very small number compared with salt.) That's why the salt in brine can penetrate an entire turkey in 24–48 hours while an acidic marinade penetrates less than the outside ¼ inch of the food.

I thank Pia Sörensen, PhD, senior preceptor in chemical engineering and applied materials with the Harvard John A. Paulson School of Engineering and Applied Sciences for informing me of the use the random walk concept and the simplified equation for calculating the rate of diffusion of ions and molecules into food. Both the concept and the equation are used in the Science and Cooking course at Harvard. In addition, Sörensen pointed out the very low concentration of protons in vinegar as an important factor in understanding why weak acids diffuse so slowly into meat.

REFERENCES

Lebert, A., and J.-D. Daudin. "Modelling the Distribution of a_w, pH and Ions in Marinated Beef Meat." *Meat Science* 97 (2014): 347–357.

Rodger, G., R. Hastings, C. Cryne, and J. Bailey. "Diffusion Properties of Salt and Acetic Acid Into Herring and Their Subsequent Effect on the Muscle Tissue." *Journal of Food Science* 49, no. 3 (1984): 714–720.

Tender cuts of beef, like tenderloin, should be cooked to an internal temperature of 120°F–125°F, or 49°C–52°C (it reaches the well-done stage at 130°F/54°C). As large cuts of beef are sterile inside, they can be cooked to relatively low temperatures and still be perfectly safe to eat as long as the surface becomes very hot. Recently, the USDA lowered the recommended safe temperature for whole cuts of pork to 145°F (60°C). Tough shoulder cuts of beef and pork containing a lot of connective tissue should be cooked in a low oven (325°F/163°C) for at least several hours to an internal temperature of 190°F (88°C) to break down the collagen protein in that connective tissue to gelatin. Collagen is a strong protein composed of three chains of gelatin wound into a triple helical structure held together by numerous hydrogen bonds. Collagen does not start to slowly unwind until 140°F (40°C), and, ultimately, it takes several hours, often as many as 6 hours, at higher temperatures to fully disentangle to gelatin. Gelatin is a unique protein that is capable of binding ten times its weight of water, so even though tough cuts of meat like barbequed pork are cooked to 190°F (88°C), they will remain relatively moist. Because chicken and turkey may contain harmful *Salmonella* bacteria within their muscle tissue, they must be cooked to 160°F (71°C) throughout to ensure safety. Chicken and turkey, as well as ground beef and ground pork, require cooking to an internal temperature of 160°F (71°C).

The Future of Cooking Science

Over the last several years, the emerging field of *culinary medicine* has been added to the realm of medical education. Culinary medicine has been defined as "a new evidence-based field in medicine that blends the art of food and cooking with the science of medicine." Its objective is "to attempt to empower the patient to care for herself or himself safely, effectively, and happily with food and beverage as a primary care technique." A number of medical schools, including Tulane, Harvard, and Northwestern, have implemented classes and certificate programs in culinary medicine to instruct doctors and patients on how to use food and cooking to improve and maintain their health at home. But stop and think about how the field of culinary medicine has been defined above. That definition refers to cooking as an art and medicine as a science! Culinary medicine will not achieve its stated objective until cooking is understood to be a science like medicine. How can sound advice be given to the patient about what foods are healthy to eat without thoroughly understanding how different methods of cooking those foods will impact their nutritional quality! If a recommendation is made to eat more healthy vegetables, then a recommendation must also be made on the best way to cook those vegetables to optimize the level of specific vitamins and key phytonutrients. Steaming may be best to optimize the glucosinolates in broccoli, but sautéing in olive oil is best for releasing the lycopenes in tomatoes. Cooking has moved well beyond the world of art and into the world of science, just like medicine did decades ago. Cooking science offers the potential to increase the quality of human life.

The science of cooking is still in its infancy, yet in the past 20 years, it has caught the attention of the cooking world and is poised for explosive growth. It's not going to explode because it teaches us how to cook a tender, juicy steak, or how to make the perfect pie dough, or how to kick up the flavor of your favorite dish. It's going to explode because it teaches us how to optimize the nutritional quality of the food we eat by minimizing the loss of essential nutrients like vitamins and minerals and enhancing the other nutrients like beneficial antioxidants, prebiotics, and phytonutrients.

The science-driven changes in the way we cook will help reduce the risk of developing chronic diseases such as heart disease, stroke, obesity, type 2 diabetes, dementia, and many forms of cancer. Cooking science will enhance the quality and joy of life. Cooking will be seen no longer as just an art but as a perfect blend of art and science, creating simple dishes that are delicious to eat and good for our health.

Bibliography

Have you noticed there are literally dozens of "kitchen myths" related to cooking? So many that there is even a recently published book on the subject. A myth is a falsehood, a fiction, or a half-truth that has been devised to explain why something happens, but in reality, it is a fabricated answer that is untrue. The classic example is the myth that searing meat seals in juices; it was proved false in 1930 but still persists even today. A more recent example is the myth that Arborio rice, used for the preparation of risotto, forms a thick, creamy sauce because the starch in Arborio rice contains virtually no amylose (starch composed of only amylopectin is known as waxy starch). This explanation makes a good story—except it's not true. A search of the scientific literature reveals that the starch in Arborio rice contains 17 percent amylose on a dry basis, while the starch in long-grain rice contains 24 percent amylose and waxy glutinous rice contains less than 4 percent amylose (based on research conducted at the Guelph Food Research Center and published in *Cereal Chemistry* in 2010). The point of this discussion is that "scientific" explanations of why certain things happen in cooking should be questioned if they are not backed up by the scientific literature and/or well-designed kitchen tests. Too many cooking science explanations are fabricated to support the observations rather than being proved by scientific experiments. In my view, the bibliography that follows is the most important part of this book, as it reveals the sources used to support the statements made throughout the text. Fortunately, my position with Harvard University gives me access to these critically important sources.

Preface

Barki, R., J. Rosell, R. Blasco, and A. Gopher. "Fire for a Reason: Barbeque at Middle Pleistocene Qesem Cave, Israel." *Current Anthropology* 58, supp. S16 (2017): S314–S327.

Bentley, G. E., Jr. *The Stranger from Paradise—A Biography of William Blake*, 309. New Haven, CT: Yale University Press, 2001.

Bronowski, J. *The Ascent of Man*, 351. Boston: Little Brown, 1973.

Wrangham, R. W. *Catching Fire: How Cooking Made Us Human*. New York: Simon and Schuster, 2009.

Wrangham, R. W. "Control of Fire in the Paleolithic: Evaluating the Cooking Hypothesis." *Current Anthropology* 58, supp. 16 (2017): S303–S313.

1. The Evolution of Cooking

Barki, R., J. Rosell, R. Blasco, and A. Gopher. "Fire for a Reason: Barbeque at Middle Pleistocene Qesem Cave, Israel." *Current Anthropology* 58, supp. S16 (2017): S314–S327.

Breslin, P. A. S. "An Evolutionary Perspective on Food and Human Taste." *Current Biology* 23 (2013): R409–R418.

Bronowski, J. *The Ascent of Man*, 62. Boston: Little Brown, 1973.

Burton, F. D. *Fire: The Spark That Ignited Human Evolution*. Albuquerque: University of New Mexico Press, 2009.

Carmody, R. N., G. S. Weintraub, and R. W. Wrangham. "Energetic Consequences of Thermal and Nonthermal Food Processing." *Proceedings of the National Academy of Sciences* 108, no. 48 (2011): 19199–19203.

Craig, O. E., H. Saul, A. Lucquin, Y. Nishida, K. Tache, L. Clarke, A. Thompson et al. "Earliest Evidence for the Use of Pottery." *Nature* 496 (2013): 351–354.

Crosby, G. "Super-tasters and Non-tasters: Is It Better to Be Average?" *The Nutrition Source* (Harvard T. H. Chan School of Public Health), May 31, 2016. https://www.hsph.harvard.edu/nutritionsource/2016/05/31/super-tasters-non-tasters-is-it-better-to-be-average.

Editors of America's Test Kitchen and G. Crosby. *Cook's Science*, 44. Brookline, MA: America's Test Kitchen, 2016.

Fernandez-Armesto, F. *Near a Thousand Tables—A History of Food*, 11. New York: Free Press, 2002.

Gowlett, J. A. J., and R. W. Wrangham. "Earliest Fire in Africa: Towards the Convergence of Archeological Evidence and the Cooking Hypothesis." *Azania: Archeological Research in Africa* 48, no. 1 (2013): 5–30.

Hoffmann, D. L., D. E. Angelucci, V. Villaverde, J. Zapata, and J. Zilhão. "U-Th Dating of Carbonate Crusts Reveals Neanderthal Origin of Iberian Cave Art." *Science Advances* 4 (2018): eaar5255.

Hoover, K. C. "Smell with Inspiration: The Evolutionary Significance of Olfaction." *Yearbook of Physical Anthropology* 53 (2010): 63–74.

Leonard, W. R., and M. L. Robertson. "Evolutionary Perspectives on Human Nutrition: The Influence of Brain and Body Size on Diet and Metabolism." *American Journal of Human Biology* 6 (1994): 77–88.

Pelchat, M. L., A. Johnson, R. Chan, J. Valdez, and J. D. Ragland. “Images of Desire: Food-Craving Activation During fMRI.” *Neuroimage* 23 (2004): 1486–1493.
Prescott, J. *Taste Matters: Why We Like the Foods We Do*. London: Reaktion Books, 2012.
Shepherd, G. M. *Neurogastronomy: How the Brain Creates Flavor and Why It Matters*. New York: Columbia University Press, 2012.
Smith, B. D. *The Emergence of Agriculture* (New York: Scientific American Library, 1995).
Speth, J. D. “When Did Humans Learn to Boil?” *PaleoAnthropology* (2015): 54–67.
Svoboda, J., M. Kralik, V. Culikova, and S. Hladilova. “Pavlov VI: An Upper Paleolithic Living Unit.” *Antiquity* 83, no. 320 (2009): 282–295.
Watford, M., and A. G. Goodridge. “Regulation of Fuel Utilization.” In M. H. Stipanuk, *Biochemical and Physiological Aspects of Human Nutrition*. Philadelphia: Saunders, 2000.
Wrangham, R. W. *Catching Fire: How Cooking Made Us Human*. New York: Simon and Schuster, 2009.

2. The Dawn of Agriculture Revolutionizes Cooking

Albala, K. *Three World Cuisines: Italian, Mexican, Chinese*, 86–88. Plymouth, UK: AltaMira Press, 2012.
Bronowski, J. *The Ascent of Man*, 70, 170. Boston: Little Brown, 1973.
Coultate, T. *Food: The Chemistry of Its Components*, 116–122. 6th ed. Cambridge, UK: Royal Society of Chemistry, 2016.
Crosby, G. “Do Cooking Oils Present a Health Risk?” *Food Technology* 72, no. 5 (2018): 50–56.
Harlan, J. R. “A Wild Wheat Harvest in Turkey.” *Archaeology* 20, no. 3 (1967): 197–201.
Hawkes, J. *The Atlas of Early Man*. New York: St. Martin’s Press, 1976.
Leicester, H. M., and H. S. Klickstein. *A Source Book in Chemistry: 1490–1900*, 1–2. Cambridge, MA: Harvard University Press, 1952.
Lu, H., Y. Li, J. Zhang, X. Yang, M. Ye, Q. Li, C. Wang, and N. Wu. “Component and Simulation of the 4000-Year-Old Noodles Excavated from the Archaeological Site of Lajia in Qinghai, China.” *Chinese Science Bulletin* 59 (2014): 5136–5152.
Müller, N. S., A. Hein, V. Kilikoglou, and P. M. Day. “Bronze Age Cooking Pots: Thermal Properties and Cooking Methods.” *Prehistoires Mediterranéennes* 4 (2013): 1–11.
Resnick, R., and D. Halliday. “Heat and the First Law of Thermodynamics.” Chap. 22 in *Physics for Students of Science and Engineering*, 466–488. New York: Wiley, 1960.
Smith, B. D. *The Emergence of Agriculture*. New York: Scientific American Library, 1995.
Symons, M. *A History of Cooks and Cooking*, 67–69, 77–78. Chicago: University of Illinois Press, 2000.
Yin-Fei Lo, E. *Mastering the Art of Chinese Cooking*, 60–61. San Francisco: Chronicle Books, 2009.

3. Early Science Inspires Creativity in Cooking

Albala, K. *Three World Cuisines: Italian, Mexican, Chinese*, 60. Plymouth, UK: AltaMira Press, 2012.

Bronowski, J. *The Ascent of Man*, 146–149. Boston: Little Brown, 1973.

Brown, G. I. *Count Rumford: Scientist, Soldier, Statesman, Spy, the Extraordinary Life of a Scientific Genius*. Gloucestershire, UK: Sutton, 1999.

Elman, B. *A Cultural History of Modern Science in China*. Cambridge, MA: Harvard University Press, 2006.

Leicester, H. M., and H. S. Klickstein. *A Source Book in Chemistry: 1490–1900*, 7, 33–38, 101, 112, 180. Cambridge, MA: Harvard University Press, 1952.

McGee, H. *On Food and Cooking: The Science and Lore of the Kitchen*, 586. New York: Scribner, 2004.

Resnick, R., and D. Halliday. *Physics for Students of Science and Engineering*, 465–466. New York: Wiley, 1960.

Waley-Cohen, J. "Celebrated Cooks of China's Past." *Flavor and Fortune* 14, no. 4 (2007): 5–7.

4. The Art of Cooking Embraces the Science of Atoms

Appert, N. *The Art of Preserving All Kinds of Animal and Vegetable Substances for Several Years* (Translated from the French). 2nd ed. London: Cox and Baylis, 1812.

Beattie, O., and J. Geiger. *Frozen in Time: The Fate of the Franklin Expedition*. Vancouver: Greystone Books, 1987.

Bentley, R. "The Nose as a Stereochemist: Enantiomers and Odor." *Chemical Reviews* 106 (2006): 4099–4112.

Brock, W. H. *Justus von Liebig: The Chemical Gatekeeper*. Cambridge: Cambridge University Press, 1997.

Bronowski, J. *The Ascent of Man*, 153. Boston: Little Brown, 1973.

Cookman, S. *Ice Blink: The Tragic Fate of Sir John Franklin's Lost Polar Expedition*. New York: Wiley, 2000.

Ferguson, P. P. "Writing Out of the Kitchen: Carême and the Invention of French Cuisine." *Gastronomica: The Journal of Food and Culture* 3 no. 3 (2003): 40–51.

Kellogg, E. E. *Science in the Kitchen*. Battle Creek, MI: Health Publishing, 1892.

Leicester, H. M., and H. S. Klickstein. *A Source Book in Chemistry: 1490–1900*, 208–220, 374–379. Cambridge, MA: Harvard University Press, 1952.

5. Modern Science Transforms the Art of Cooking

Baldwin, D. "Sous Vide Cooking: A Review." *International Journal of Gastronomy and Food Science* 1 (2012): 15–30.

Cassi, D. "Science and Cooking: The Era of Molecular Cuisine." *European Molecular Biology Organization Reports* 12, no. 3 (2011): 191–196.

Christlbauer, M., and P. Schieberle. "Evaluation of the Key Aroma Compounds in Beef and Pork Vegetable Gravies a la Chef by Stable Isotope Dilution Assays and Aroma Recombination Experiments." *Journal of Agricultural and Food Chemistry* 59 (2011): 13122–13130.

Crosby, G. "The Top Ten Breakthroughs in Food Science in the Past 75 Years." *Food Technology* 69, no. 7 (2015): 120.

Editors of America's Test Kitchen and G. Crosby. *Cook's Science*, 192–195. Brookline, MA: America's Test Kitchen, 2016.

Eertmans, A., F. Baeyens, and O. Van den Bergh. "Food Likes and Their Relative Importance in Human Eating Behavior: Review and Preliminary Suggestions for Health Promotion." *Health Education Research* 16 (2001): 443–456.

Felder, D., D. Burns, and D. Chang. "Defining Microbial Terroir: The Use of Native Fungi for the Study of Traditional Fermentative Processes." *International Journal of Gastronomy and Food Science* 2 (2012): 64–69.

Hodge, J. "Dehydrated Foods: Chemistry of Browning Reactions in Model Systems." *Journal of Agricultural and Food Chemistry* 1, no. 15 (1953): 928–943.

Rosler, M. *The Art of Cooking: A Dialogue Between Julia Child and Craig Claiborne*. Minneapolis: University of Minnesota Press, 2016. https://www.classicscookbooks.com/blogs/notes/36967812-martha-rosler-the-art-of-cooking-a-dialogue-between-julia-child-and-craig-claiborne.

Keller, T. *Under Pressure: Cooking Sous Vide*. New York: Artisan, 2008.

Reineccius, G. *Flavor Chemistry and Technology*. 2nd ed. Boca Raton, FL: CRC Press, 2006.

Risch, S. J., and C-T. Ho. *Flavor Chemistry: Industrial and Academic Research*. Washington, DC: American Chemical Society, 2000.

Shepherd, G. M. "Smell Images and the Flavour System in the Human Brain." *Nature* 444 (2006): 316–321.

Shepherd, G. M. *Neurogastronomy: How the Brain Creates Flavor and Why It Matters*. New York: Columbia University Press, 2012.

Teranishi, R., E. M. Wick, and I. Hornstein, eds. *Flavor Chemistry: Thirty Years of Progress*. New York: Springer Science, 1999.

This, H. *Molecular Gastronomy: Exploring the Science of Flavor*. New York: Columbia University Press, 2006.

Turbek, A. B. *The Taste of Place: A Cultural Journey Into Terroir*. Berkeley: University of California Press, 2008.

van den Linden, E., D. McClements, and J. Ubbink. "Molecular Gastronomy: A Food Fad or an Interface for Science-Based Cooking?" *Food Biophysics* 3 (2009): 246–254.

Zamaora, R., and F. J. Hidalgo. "Coordinate Contribution of Lipid Oxidation and Maillard Reaction to the Nonenzymatic Food Browning." *Critical Reviews in Food Science and Nutrition* 45 (2005): 49–59.

6. Cooking Science Catches Fire!

Baldwin, D. "Sous Vide Cooking: A Review." *International Journal of Gastronomy and Food Science* 1 (2012): 15–30.

Coultate, T. *Food: The Chemistry of Its Components*, 360. 6th ed. Cambridge: Royal Society of Chemistry, 2016.

Crosby, G. "Pairing Cooking Science with Nutrition," 2014. https://www.youtube.com/watch?v=3vNE3eiU_Sw.

Editors of America's Test Kitchen and G. Crosby. *The Science of Good Cooking*. Brookline, MA: America's Test Kitchen, 2012.

Fabbri, A. D. T., and G. Crosby. "A Review of the Impact of Preparation and Cooking on the Nutritional Quality of Vegetables and Legumes." *International Journal of Gastronomy and Food Science* 3 (2016): 2–11. https://www.sciencedirect.com/science/article/pii/S1878450X15000207#!.

Fielding, J., K. Rowley, P. Cooper, and K. O'Dea. "Increases in Plasma Lycopene Concentration After Consumption of Tomatoes Cooked with Olive Oil." *Asia Pacific Journal of Clinical Nutrition* 14, no. 2 (2005): 131–136.

Gartner, C., W. Stahl, and H. Sies. "Lycopene Is More Bioactive from Tomato Paste Than from Fresh Tomatoes." *American Journal of Clinical Nutrition* 66 (1997): 116–122.

Giovannucci, E. "A Review of Epidemiological Studies of Tomatoes, Lycopene, and Prostate Cancer." *Society for Experimental Biology and Medicine* 227 (2002): 852–859.

Harris, R., and E. Karmas, eds. *Nutritional Evaluation of Food Processing*. 2nd ed. Westport, CT: AVI Publishing, 1975.

Kahlon, T., R. Miczarek, and M-C. Chiu. "*In Vitro* Bile Acid Binding of Mustard Greens, Kale, Broccoli, Cabbage and Green Bell Pepper Improves with Sautéing Compared with Raw or Other Methods of Preparation." *Food and Nutrition Sciences* 3 (2012): 951–958.

Kon, S. "Effect of Soaking Temperature on Cooking and Nutritional Quality of Beans." *Journal of Food Science* 44 (1979): 1329–1340.

McGee, H. *On Food and Cooking: The Science and Lore of the Kitchen*. New York: Scribner, 2004.

McNaughton, S., and G. Marks. "Development of a Food Composition Database for the Estimation of Dietary Intakes of Glucosinolates, the Biologically Active Constituents of Cruciferous Vegetables." *British Journal of Nutrition* 90 (2003): 687–697.

Miglio, C., E. Chiavaro, A. Visconti, V. Fogliano, and N. Pellegrini. "Effects of Different Cooking Methods on Nutritional and Physicochemical Characteristics of Selected Vegetables." *Journal of Agricultural and Food Chemistry* 56 (2008): 139–147.

Palermo, M., N. Pellegrini, and V. Fogliano. "The Effect of Cooking on the Phytochemical Content of Vegetables." *Journal of the Science of Food and Agriculture* 94 (2014): 1057–1070.

Pellegrini, N. Personal communication with the author, August 2018.

Perla, V., D. G. Holm, and S. S. Jayanty. "Effects of Cooking Methods on Polyphenols, Pigments and Antioxidant Activity in Potato Tubers." *LWT—Food Science and Technology* 45 (2012): 161–171.

Rickman, J., D. Barrett, and C. Bruhn. "Nutritional Comparison of Fresh, Frozen, and Canned Fruits and Vegetables. Part 1. Vitamins C and B and Phenolic Compounds." *Journal of the Science of Food and Agriculture* 87 (2007): 930–944.

Stipanuk, M. *Biochemical and Physiological Aspects of Human Nutrition*. Philadelphia: Saunders, 2000.

Verhoven, D., R. Goldborm, G. van Popple, H. Verhagen, and P. van den Brandt. "Epidemiological Studies on Brassica Vegetables and Cancer Risk." *Cancer Epidemiology, Biomarkers and Prevention* 5 (1996): 733–748.

Willett, W. *Eat, Drink, and Be Healthy: The Harvard Medical School Guide to Healthy Eating*. New York: Simon & Schuster, 2017.

7. The Good, the Bad, and the Future of Cooking Science

Cahill, L., A. Pan, S. Chiuve, W. Willett, F. Hu, and E. Rimm. "Fried-Food Consumption and Risk of Type 2 Diabetes and Coronary Heart Disease: A Prospective Study in 2 Cohorts of US Women and Men." *American Journal of Clinical Nutrition* 100 (2014): 667–675.

Crosby, G. "Do Cooking Oils Present a Health Risk?" *Food Technology* 72, no. 5 (2018): 50–56.

Dunford, N. *Food Technology Fact Sheet No. 126: Deep Fat Frying Basics for Food Services*. Stillwater: Oklahoma Cooperative Extension Service, 2017.

Duyff, R., J. Mount, and J. Jones. "Sodium Reduction in Canned Beans After Draining, Rinsing." *Journal of Culinary Science and Technology* 9 (2011): 106–112.

Estruch, R., E. Ros, J. Salas-Salvadó, M-I. Covas, D. Corella, F. Arós, E. Gómez-Gracia et al. "Primary Prevention of Cardiovascular Disease with a Mediterranean Diet." *New England Journal of Medicine* 368 (2013): 1279–1290.

Fabbri, A., R. Schacht, and G. Crosby. "Evaluation of Resistant Starch Content of Cooked Black Beans, Pinto Beans, and Chickpeas." *NFS Journal* 3 (2016): 8–12.

Fuentes-Zaragoza, E., M. Riquelme-Navarrete, E. Sanchez-Zapata, and J. Perez-Alvarez. "Resistant Starch as a Functional Ingredient: A Review." *Food Research International* 43, no. 4 (2010): 931–942.

Gadiraju, T., Y. Patel, J. Gaziano, and L. Djouss. "Fried Food Consumption and Cardiovascular Health: A Review of Current Evidence." *Nutrients* 7 (2015): 8424–8430.

Gil-Humanes, J., F. Piston, R. Altimirano-Fortoul, A. Real, I. Comino, C. Sousa, C. Rosell, and F. Barro. "Reduced-Gliadin Wheat Bread: An Alternative to the Gluten-Free Diet for Consumers Suffering Gluten-Related Pathologies." *PLOS ONE* 9, no. 3 (2014): e90898.

Katragadda, H. R., A. Fullana, S. Sidhu, and A. A. Carbonell-Barrachina. "Emissions of Volatile Aldehydes from Heated Cooking Oils." *Food Chemistry* 120 (2010): 59–65.

Kon, S. "Effect of Soaking Temperature on Cooking and Nutritional Quality of Beans." *Journal of Food Science* 44 (1979): 1329–1340.

La Puma, J. "What Is Culinary Medicine and What Does It Do?" *Population Health Management* 19, no. 1 (2016): 1–3.

Monnier, V. "Dietary Advanced Lipoxidation Products as Risk Factors for Human Health—A Call for Data." *Molecular Nutrition and Food Research* 51 (2007): 1091–1093.

Murphy, M., J. Douglass, and A. Birkett. "Resistant Starch Intakes in the United States." *Journal of the American Dietetic Association* 108 (2008): 67–78.

Sebastian, A., S. Ghazani, and A. Maragoni. "Quality and Safety of Frying Oils Used in Restaurants." *Food Research International* 64 (2014): 420–423.

Vaziri, N., S-M. Liu, W. L. Lau, M. Khazaeli, S. Nazertehrani, S. H. Farzaneh, D. A. Kieffer et al. "High Amylose Resistant Starch Diet Ameliorates Oxidative Stress, Inflammation, and Progression of Chronic Kidney Disease." *PLOS ONE* 9, no. 12 (2014): 114881–114895.

Willett, W. *Eat, Drink, and Be Healthy: The Harvard Medical School Guide to Healthy Eating*. New York: Simon & Schuster, 2017.

Index

Page numbers in *italics* indicate figures or tables.

Arts and Traditions of the Table: Perspectives on Culinary History
Albert Sonnenfeld, Series Editor

Salt: Grain of Life, Pierre Laszlo, translated by Mary Beth Mader
Culture of the Fork, Giovanni Rebora, translated by Albert Sonnenfeld
French Gastronomy: The History and Geography of a Passion, Jean-Robert Pitte, translated by Jody Gladding
Pasta: The Story of a Universal Food, Silvano Serventi and Françoise Sabban, translated by Antony Shugar
Slow Food: The Case for Taste, Carlo Petrini, translated by William McCuaig
Italian Cuisine: A Cultural History, Alberto Capatti and Massimo Montanari, translated by Áine O'Healy
British Food: An Extraordinary Thousand Years of History, Colin Spencer
A Revolution in Eating: How the Quest for Food Shaped America, James E. McWilliams
Sacred Cow, Mad Cow: A History of Food Fears, Madeleine Ferrières, translated by Jody Gladding
Molecular Gastronomy: Exploring the Science of Flavor, Hervé This, translated by M. B. DeBevoise
Food Is Culture, Massimo Montanari, translated by Albert Sonnenfeld
Kitchen Mysteries: Revealing the Science of Cooking, Hervé This, translated by Jody Gladding
Hog and Hominy: Soul Food from Africa to America, Frederick Douglass Opie
Gastropolis: Food and New York City, edited by Annie Hauck-Lawson and Jonathan Deutsch
Building a Meal: From Molecular Gastronomy to Culinary Constructivism, Hervé This, translated by M. B. DeBevoise
Eating History: Thirty Turning Points in the Making of American Cuisine, Andrew F. Smith
The Science of the Oven, Hervé This, translated by Jody Gladding
Pomodoro! A History of the Tomato in Italy, David Gentilcore
Cheese, Pears, and History in a Proverb, Massimo Montanari, translated by Beth Archer Brombert
Food and Faith in Christian Culture, edited by Ken Albala and Trudy Eden
The Kitchen as Laboratory: Reflections on the Science of Food and Cooking, edited by César Vega, Job Ubbink, and Erik van der Linden
Creamy and Crunchy: An Informal History of Peanut Butter, the All-American Food, Jon Krampner
Let the Meatballs Rest: And Other Stories About Food and Culture, Massimo Montanari, translated by Beth Archer Brombert
The Secret Financial Life of Food: From Commodities Markets to Supermarkets, Kara Newman
Drinking History: Fifteen Turning Points in the Making of American Beverages, Andrew F. Smith
Italian Identity in the Kitchen, or Food and the Nation, Massimo Montanari, translated by Beth Archer Brombert
Fashioning Appetite: Restaurants and the Making of Modern Identity, Joanne Finkelstein

The Land of the Five Flavors: A Cultural History of Chinese Cuisine, Thomas O. Höllmann, translated by Karen Margolis

The Insect Cookbook: Food for a Sustainable Planet, Arnold van Huis, Henk van Gurp, and Marcel Dicke, translated by Françoise Takken-Kaminker and Diane Blumenfeld-Schaap

Religion, Food, and Eating in North America, edited by Benjamin E. Zeller, Marie W. Dallam, Reid L. Neilson, and Nora L. Rubel

Umami: Unlocking the Secrets of the Fifth Taste, Ole G. Mouritsen and Klavs Styrbæk, translated by Mariela Johansen and designed by Jonas Drotner Mouritsen

The Winemaker's Hand: Conversations on Talent, Technique, and Terroir, Natalie Berkowitz

Chop Suey, USA: The Story of Chinese Food in America, Yong Chen

Note-by-Note Cooking: The Future of Food, Hervé This, translated by M. B. DeBevoise

Medieval Flavors: Food, Cooking, and the Table, Massimo Montanari, translated by Beth Archer Brombert

Another Person's Poison: A History of Food Allergy, Matthew Smith

Taste as Experience: The Philosophy and Aesthetics of Food, Nicola Perullo

Kosher USA: How Coke Became Kosher and Other Tales of Modern Food, Roger Horowitz

Chow Chop Suey: Food and the Chinese American Journey, Anne Mendelson

Mouthfeel: How Texture Makes Taste, Ole G. Mouritsen and Klavs Styrbæk, translated by Mariela Johansen

Garden Variety: The American Tomato from Corporate to Heirloom, John Hoenig